CONFINED
WITHDRAWN

Not For Loan

# TIMBERS of the WORLD

# TIMBERS of the WORLD

TRADA/The Construction Press

The Construction Press Ltd.,
Lancaster, England

A subsidiary company of Longman Group Ltd., London
Associated companies, branches and representatives
throughout the world.

Published in the United States of America by
Longman Inc., New York.

© Timber Research and Development Association, 1979.

Originally issued as a series of separate publications by the
Timber Research and Development Association.

First published in this format, 1979.

ISBN  0  860958  36  1

Printed in Great Britain by the Pitman Press, Bath

# CONTENTS

# PREFACE

*In 1945 the Timber Development Association, as it was then known, revised a Red Booklet entitled 'Timbers of British West Africa' and republished it as 'Timbers of West Africa'. At that time only a few African timbers were well known to the trade, mainly from the west coast, and the need for information regarding what were then lesser known timbers was great, and for many years the original content and geographical coverage of the booklet were considered adequate. It was revised in 1968 and again in 1972, by the Timber Research and Development Association, and together with its companion booklets, 'Timbers of South East Asia', and 'Timbers of South America', provided a useful service to trade and industry by detailing the practical characteristics of many commercial timber species from these areas.*

*Timber as a major raw material came to be in greater demand and, in consequence, there was a greater need for a wide knowledge of the world's timber resources. With this in mind, the Timber Research and Development Association decided to publish a series of booklets giving a wider and more adequate account of the commercial timbers of the world. This series of booklets is now brought together to form the present 2–Volume Timbers of the World.*

## DURABILITY OF TIMBERS

*Durability, or resistance to decay is important when woods are selected for certain uses where the conditions are favourable for decay to occur. Sapwood is nearly always perishable in these conditions, but generally more permeable than heartwood, consequently it should not be used in exposed situations without preservative treatment. Heartwood varies in its natural resistance to decay according to the species and the amount of decay inhibiting substances contained in the wood.*

*The various grades of durability mentioned in the text are those resulting from exposure tests carried out in the United Kingdom and, accordingly, are approximate values applicable to areas with similar climate.*

*The tests refer to all-heartwood stakes of 50mm x 50mm section driven in the ground. The five durability grades are defined as follows:*

| | |
|---|---|
| Perishable | Less than 5 years when in contact with the ground. |
| Non-durable | 5–10 years when in contact with the ground. |
| Moderately durable | 10–15 years when in contact with the ground. |
| Durable | 15–25 years when in contact with the ground. |
| Very durable | More than 25 years when in contact with the ground. |

## LAYOUT OF THE BOOK

*For each geographical area (that is, in each chapter) the timbers are placed in two groups, ie hardwoods and softwoods, and are arranged in alphabetical order of their common names; these names are, wherever possible, the standardised ones given in BS 881 and 589: 1974: 'Nomenclature of commercial timbers, including sources of supply'.*

*Following the notes on individual timbers each chapter has a guide listing those timbers suitable for specific purposes and a section dealing with the amenability of heartwood to preservative treatment. The book ends with references and an index.*

# 1
# AFRICA

Mediterranean Sea

MOROCCO

ALGERIA

LIBYA

EGYPT

SPANISH
SAHARA

Red Sea

MAURITANIA

MALI
(SOUDAN)

NIGER

CHAD

SUDAN

DJIBOUTI

SENEGAL

GAMBIA

GUINEA

GUINEA-BISSAU

SIERRA LEONE

LIBERIA

GHANA

TOGO

BENIN

NIGERIA

CAMEROON
REPUBLIC

CENTRAL
AFRICAN REPUBLIC

ETHIOPIA

SOMALIA

UGANDA

Gulf of Guinea

GABON

CONGO(DEM REP)

ZAIRE

TANZANIA

Indian Ocean

ANGOLA

ZAMBIA

MOZAMBIQUE

MALAGASY

Atlantic Ocean

NAMIBIA

RHODESIA

BOTSWANA

SOUTH AFRICA

# INTRODUCTION

*In this chapter Africa is considered as a whole, but with a few minor omissions as to geographical forest areas.*

*In the north of the continent, to a great extent, the Atlas mountain regions form part of the south European botanical zone and accordingly species of poplar, lime and oak, which occur in Morocco, for example, are not included in this chapter but are dealt with more appropriately under 'Europe' in Volume 2.*

## FORESTS OF AFRICA

*Only about 10 per cent of the total land area of the African continent is covered with forests; tropical hardwoods predominate, comprising about 96 per cent of the forests, with temperate hardwoods accounting for about 3 per cent, and softwoods only about 1 per cent.*

### North Africa
*Forest vegetation has largely disappeared from Algeria and Tunisia, but there is still some forest land in Morocco where the trees have a close affinity with those of Spain and Portugal.*

### West and East Africa
*The great forest areas are the tropical forests which cover the coastal belt of West Africa, from Guinea to Gabon, and which extend into the basin of the Congo. To the east of the Great Rift Valley, Uganda, Kenya and Tanzania, all have large forest areas, as does also Angola in the south-west , and Mozambique in the south-east.*

*Two types of forest occur; the mature, or 'high' forests are typical of the tropical or rain forests where rainfall exceeds 1500mm per year without prolonged dry period. The heaviest rainfall occurs in the coastal areas of the west, where it averages about 4000mm per year. An open park-like forest occurs in regions where the rainfall amounts to 750mm to 1000mm per year. Large areas of these dry-savannah forests occur in East Africa, Angola and Rhodesia, particularly, but similar fringe-forests occur in West and Central Africa.*

*Forest reserves, owned by regional governments or local authorities of the various countries, are administered under strict control.*

1

*Nigeria for example has some 2 million hectares of high forest reserves, either under exploitation or leased to timber concessionaires, who hold exclusive felling rights. There is also a large area of roughly 6 million hectares of savannah forest reserves in which usable but scattered quantities of timber trees occur. In addition to the reserves there are large areas of forest not so strictly controlled, which produce about 50 per cent of Nigeria's total volume of timber.*

## South Africa

*South Africa is poorly endowed with natural forests capable of producing good timber trees, because well over 73 per cent of the surface of the Republic has an annual rainfall of less than 635mm and summer temperatures are high. If the scattered cedar trees on the Cedarberg of Clanwilliam and on the Baviaanskloof Mountains, and the savannah forests of the Transvaal and Natal low-veld are disregarded — because they are not true forests — it is possible to say briefly that indigenous timber forests occur as a non-continuous belt in the region of high rainfall extending east and north-east from Table Mountain to the North-Eastern Transvaal.*

*The only really extensive wooded areas in South Africa are in the George-Knysna region, where there are more than 48,000 hectares of timber forest on the narrow plateau between the ocean and the Duteniqua and Tsitsikamma Ranges.*

*Because the indigenous tree species were, especially on account of their slow growth, unsuitable for afforestation, South Africa has imported trees for this purpose, and state owned plantations have progressivly been established, concentrating on the cultivation of Eucalyptus species from Australia, and various pines from the Mediterranean, Mexico and the Southern USA. South Africa is now more self-sufficient in respect of timber and timber products; the main industries, mining, fruit farming and wine production, absorb much indigenous timber for waggon building, railway sleepers, boxes, crates, etc.,while timber from maturing introduced tree species has improved the range of woods required for consumer goods and for building purposes. Exports of timber and allied products from South Africa consist mainly of flooring timber, rayon and paper pulp, fibre-board, chipboard, plywood and matches.*

*Consumption of timber in South Africa for all purposes was 5.3 million cubic metres in 1972, and the demand is expected to grow to 23 million cubic metres by the year 2000.*

# PART I   HARDWOODS

## ABURA

*Mitragyna ciliata* Andrew & Pellgr.          Family: Rubiaceae

**Other names**
subaha (Ghana); bahia (French W Africa); elilom (Cameroons); elelome (Gabon); maza, voukou, vuku (Zaire); mujiwa, mushiwa.

**Distribution**
Occurs in West Tropical Africa from Sierra Leone through Liberia and other coastal countries to the Cameroons and Gabon. It is probable that most of the wood exported is *M. ciliata* from the rain forests, since the botanically associated *M. stipulosa* grows outside this zone.

**The tree**
Abura attains a height of 30m to 40m with a diameter of 1.0m to 1.2m. The tree is free of buttresses, and the bole is straight and cylindrical.

**The timber**
Pale, reddish-brown to light brown, sapwood wide but not differentiated from heartwood. Grain straight to interlocked, sometimes spiral grain present. Texture fairly fine and very even, sometimes figured like Canadian birch. Rather soft, weight varies from 480 to 640 kg/m$^3$ dried (average 580 kg/m$^3$). Fairly resistant to acids.

**Drying**
Air and kiln dries well and with little degrade. Very stable when dried.

**Strength**
A timber with medium strength properties, resembling those of common elm.

3

## Durability
Not very resistant to decay.

## Working qualities
Works well and cleanly with hand and machine tools; takes a good finish; stains, paints, and polishes well. Holds nails and screws, and glues well. One of the best West African timbers for small mouldings. Some logs may have 'spongy heart', and are then difficult to work.

## Uses
Interior joinery, fittings, cabinets, mouldings and floors, turnery, plywood and interior doors. Owing to its resistance to acids it is valuable for battery and accumulator boxes. It is a first-class general utility wood.


# AFARA

*Terminalia superba*          Family: Combretaceae
Engl. & Diels.

## Other names
The timber has various names according to the part of West Africa from which it comes; the following list gives the most common names with their corresponding countries of origin.

Nigeria: white afara.
Zaire and Angola: limba clair or light limba, limba noir or dark limba, according to the colour of the heartwood.
French West Africa: limbo, chêne limbo, fraké, noyer du Mayombe, akom.
Ghana: ofram.
Liberia: limba, korina.


Some confusion has arisen regarding the difference between white afara and black afara. Actually these are two entirely different species; black afara is *Terminalia ivorensis* and it is usually (and more correctly) known as idigbo. The confusion has been greater because the descriptive adjectives 'white' and

4

'black' were thought to refer to the colour of the wood, since in the case of white afara the heartwood is often grey, streaked with black; black afara (idigbo), paradoxically, is of a uniform pale yellow colour. This anomaly is simply explained; the colours refer to the bark and not to the timber.

## Distribution
It is widely distributed in West Africa from Sierra Leone to the Cameroons.

## The tree
Afara is a very large tree, from 18m to 45m high. The wide, spreading buttresses may extend up the bole for 2.5m or more, but the bole above is straight and cylindrical with a diameter of about 1.5m. Felling is carried out above the buttress.

## The timber
Usually both the heartwood and sapwood are light yellowish-brown in colour, similar to light oak. Occasionally, however, the heartwood contains irregular greyish markings, with streaks which may be almost black (the cause of these markings is not known). Such timber is very attractive in appearance and fetches good prices, being valuable for veneer; off-centre peeling and quarter slicing give the best striping effects. This variation in colour has resulted in the timber from Zaire being divided into two types:—
1  Limba clair (or limba blanc), in which two-thirds or more of the diameter of the log is light in colour.
2  Limba noir (or limba bariolé), in which the dark-coloured heartwood is sufficient to show on the sides of squared logs.

The timber is close-grained and usually straight, but may be wavy in the grain; the latter type providing a good figure. It is of medium hardness and weighs about 560 kg/m$^3$ when dried (to 15 per cent moisture content).
In large logs the heart may be brittle, and if this is the case, it should be eliminated ('boxed out') in conversion.

## Drying
In air drying, there is a tendency for the heart or 'brash' wood to split and shake, but little trouble is experienced with close-

5

ringed outer wood; fairly thick sticks (25mm) assist in pre-
venting discoloration. Kiln drying is easy, and there is less
tendency for defects to develop.


## Strength
Complete strength tests have not been carried out, but limited
tests indicated that the timber is not very strong, and the dark
coloured wood tends to be more brittle than the light, which is
fairly resistant to shock loads.


## Durability
Afara is not resistant to decay and frequently the heartwood of
the larger trees is unsound when felled. The sapwood is liable
to blue sapstain. The logs are liable to be attacked by pin-hole
borers, which affect both heartwood and sapwood. Powder-
post beetles tend to attack the sapwood if care is not taken.


## Working qualities
The timber works easily with hand and machine tools; this
includes turning. In cases where the grain is uneven there is
some tendency to pick up in planing, but this can usually be
overcome by using a low cutting angle. Gluing presents no
difficulties, and an excellent finish can be obtained. The
timber can be stained and polished easily (if a filler is used).
Care must be taken in nailing and screwing to avoid splitting.


## Uses
The light coloured wood is suitable for face veneers for doors,
high quality plywood and furniture, and in appearance is
rather like light oak. The more greenish-grey wood is used for
core veneer, utility plywood, light construction work such as
school equipment, office desks and furniture. The black heart,
is very decorative and is suitable for veneer, panelling, furniture,
etc. Since grading, especially in Nigeria, is usually based on the
degree of pin-worm hole present, the terms 'worm free',
'almost worm free', and 'worm-holes considered' have special
significance in terms of end use. A degree of pin-worm hole
is therefore not necessarily a defect when the requirement is for
painting, or for blockboard cores.

6

# AFRORMOSIA

*Pericopsis elata* van Meeuwen.  Family: Leguminosae
Syn. *Afrormosia elata* Harms.

## Other names
kokrodua (Ghana, Ivory Coast) ; assamela (Ivory Coast).

## Distribution
Afrormosia is found in the Ivory Coast, Ghana and Zaire, and to a small extent in Nigeria.

## The tree
A large tree (except in Nigeria), it reaches a height of 45m and a diameter of 1.2m or slightly more.

## The timber
Sapwood small, about 12mm wide, slightly lighter in colour than the brownish heartwood. When first cut the heartwood is yellowish-brown, darkening on exposure to a pleasing brownish-yellow, somewhat resembling teak, but with a finer texture, and lacking the oily nature of teak.
The grain is straight to interlocked, and the wood weighs about 710 kg/m$^3$ when dried.

## Drying
Dries rather slowly, with little degrade.

## Strength
Afrormosia is superior to teak in most of its mechanical properties and is very similar to home grown beech, except in resistance to compression where it is about 20 per cent stronger.

## Durability
Very durable.

## Working qualities
The wood works well with only a slight tendency to pick up, but a cutting angle of 20° usually produces a good finish. It can be glued and polished satisfactorily, but tends to split when nailed.

7

## Uses

Afrormosia can be used as an alternative to teak for many purposes where a strong, stable, and durable wood is required. It is used for furniture, high-class joinery, flooring, boat-building, shop fitting.

It should not be used in contact with ferrous metals in wet conditions since these may corrode, and the presence of tannins in the wood can cause staining.

# AFZELIA

*Afzelia* spp.                                        Family : Leguminosae

The trade name afzelia has been proposed for all species of this genus. In practice the West African species are usually grouped together as a single commercial timber. The East African species is usually marketed separately.

The principal species producing West African afzelia are believed to be *A. bipindensis* and *A. pachyloba.*

## Other names

| | |
|---|---|
| *A. africana* Smith | ⎫ doussié (Cameroons and France) |
| *A. bipindensis* Harms | ⎬ apa, aligna (Nigeria.) |
| *A. pachyloba* Harms | ⎭ |
| *A. quanzensis* Welw | chamfuta, mussacossa (Mozambique) mbembakofi, mkora (Tanzania) |

## Distribution

Afzelia is a transition species found between the savannah forest of dry areas and the dense forests of humid regions. It occurs throughout West Africa, Uganda and parts of Tanzania.

## The tree

The West African species attain their greatest size in the moist deciduous forest, with a height of 12m to 18m and a diameter of 1.0m but the bole is relatively short, and rarely straight. In East Africa, it is found mainly in coastal, lowland, and savannah type forests, and is generally smaller, with the bole, above the buttressed base usually about 4.5m high and with a diameter of 1.0m.

## The timber
The various species of afzelia are very similar in appearance. The sapwood is pale straw-coloured and sharply defined from the light-brown heartwood; the latter often becomes dark red-brown on exposure. Mottle and other figure is frequently present. The grain is irregular and often interlocked and the texture is coarse but even. It is a hard and moderately heavy wood, weighing about 830 kg/m$^3$ when dried. Afzelia is an exceptionally stable wood, being comparable to teak in this respect.

## Drying
Afzelia species can be kiln dried satisfactorily, but slowly from the green condition. Degrade is not likely to be severe, slight distortion may occur with some fine checking and extension of shakes.

## Durability
It is very durable and is reported to be proof against termite and teredo.

## Strength
A strong timber, with strength properties comparable with those of oak.

## Working properties
Somewhat hard to work, but produces a good finish and may be polished to a very attractive appearance.

## Uses
Heavy construction work, especially dock work, bridge building, and flooring. It is a good furniture wood especially when figured; it may be cut into veneer, both for interior decoration and furniture. It can be used for school, office, and garden furniture, staircases, bank counters, laboratory benches, door and window frames, ships rails.

# AGBA

*Gossweilerodendron balsamiferum*          Family : Leguminosae
Harms

## Other names
moboron, tola, tola branca, white tola (Angola); ntola,
mutsekamambole (Zaire).
This species should not be confused with tchitola, (*Oxystigma,
Sindora* and *Oxyphyllum* spp.), sometimes known as tola, tola
manfuta, etc.

## Distribution
It is found in West Africa, mainly in the western province of
Nigeria, but also is found in Angola and Zaire.

## The tree
Agba is one of the largest trees of West Africa where it is found
in the rain forests. The trunk is cylindrical and free from buttresses; in fact, sometimes the trunk has the appearance of a
round log set on end, with no root swelling at the base. The
tree is extremely tall and often clear of branches to over 30m,
the diameter frequently being 1.5m to 2m.
The bark is thin, greyish, smooth and often showing spiral
twisting. If the trunk is wounded or notched, thick gum or
oleo-resin exudes which hardens into large lumps. There is
considerable variation in respect of gum, some logs being
practically free while others contain large quantities.

## The timber
There is little difference in colour between the sapwood and the
heartwood, the latter is slightly darker but the line of demarcation
is somewhat indefinite. The wood varies from yellowish-pink
to reddish brown (like a light coloured mahogany). Generally,
it strongly resembles mahogany in grain etc, but is less lustrous
and paler in colour. The texture is fine and the timber is fairly
hard. Particularly when freshly cut the surfaces tend to be
gummy. The weight is about 510 kg/m³ when dried.

## Drying
The timber can be dried fairly rapidly with very little tendency to
warp or split. Some gum exudation is likely to occur, especially
in pieces containing the pith, and for this reason, very high
temperatures should be avoided.

## Strength

Agba is highly resistant to crushing strains; it compares favourably with Honduras mahogany, being about half as tough again and equal in crushing strength, though somewhat less stiff. Brittleheart is often extensive, particularly in large logs. The affected timber is considerably weaker than normal wood, and careful selection is necessary where strength is a requirement.

## Durability

The timber is very resistant to decay.

## Working qualities

Agba is easy to work with most hand and machine tools; sometimes there is a slight tendency for saws to stick owing to the gumminess of the wood. An excellent finish can be obtained and the wood has good nailing, screwing and gluing properties. It produces good veneer by slicing, but the logs should not be steamed, or the veneer dried under high temperature, because the wood will gum under these conditions.

Generally, agba is similar to Honduras mahogany in working qualities, but usually milder to work. It takes stain and polish well.

## Uses

Joinery, interior and shop fittings, turnery, panelling, flooring, sills, furniture, motor bodies and coachwork, toys, veneer and plywood. The large sizes, free from defects, in which it is obtainable, make the wood valuable for many purposes.

## AKOSSIKA

*Scottellia* spp.                                    Family: Flacourtiaceae

Several species of *Scottellia* occur in west Africa including *S. coriacea* or odoko. Two further species should be mentioned, since their timber is on offer at the present time.

*S. chevalieri* produces akossika a grande feuilles, and *S. kamerunensis* produces akossika a petites feuilles, the former species being the larger tree, some 30m to 45m tall, and a diameter of 0.4m to 1.0m with a small buttress and a good cylindrical bole, while *S. kamerunensis* attains a height of 30m and a diameter of

0.6m with a long, straight bole which, however, is seldom cylindrical.

## The timber
*S. chevalieri* : Little distinction between sapwood and heartwood which is pale yellow with occasional darker streaks. Grain usually straight, occasionally interlocked; texture fine and even. The wood has an attractive silver-grain figure on quarter-sawn surfaces.

*S. kamerunensis* : Pale yellow in colour, wood lustrous, grain usually straight, texture medium.

Both species weigh between 580 and 640 kg/m³ when dry, and both are perishable.

## General characteristics
The timbers dry easily, but with a tendency to split and check, and blue-stain occurs, or is liable to occur, during kiln drying. The wood works quite well, takes a good finish, can be glued, and tends to split in nailing. *S. kamerunensis* requires more care in planing and moulding in order to avoid grain pick-up.

## Uses
Although *S. chevalieri* appears to be the better wood, both species are used locally for light construction, flooring, furniture, interior joinery, carving, etc. They produce good plywood when staining is avoided.

## ALBIZIA, WEST AFRICAN

*Albizia* spp.                                    Family: Leguminosae

The numerous species of *Albizia* show considerable variation in the properties of their timbers, and the precise identity of the commercial species is uncertain. The following brief notes summarize the main African species.

## Other names
ayinre (Nigeria) ; okuro (Ghana) ; nongo (East Africa).

## Distribution
The genus *Albizia* includes at least 30 species in Africa, but

many of these are small trees of the savannah forest. The species from which commercial timber is produced are chiefly trees of the high forest and occur from Sierra Leone through Central Africa to East Africa and Rhodesia.

## Main species

| | |
|---|---|
| *A. adianthifolia* W. F. Wight | |
| *A. ferruginea* Benth. | West Africa |
| *A. zygia* Macbride | |
| *A. grandibracteata* Taub. | East Africa |
| *A. zygia* Macbride. | |

## Characteristics

*A. ferruginea.* A large tree some 36m in height and 1.0m in diameter, with a clear bole of 9m to 12m. The colour of the heartwood varies from medium brown to dark chocolate-brown, clearly distinct from the yellowish-white sapwood which may be 50mm wide. The grain is decidedly interlocked and often irregular; the texture is coarse. It weighs 640 kg/m$^3$ dried.

A moderately hard, moderately heavy, very durable timber, with strength properties similar to oak. Fairly easy to work, although irregular interlocked grain may cause tearing in planing. It is said to cause mild irritation of the nose when sawing the dried wood.

*A. zygia.* A tree of medium size, about 27m high and a diameter of 1.0m. It is usually heavily buttressed, but the bole form is good. The heartwood is pale brown, with a pinkish tinge, while the sapwood, which is usually wide (150mm or more), is white, yellowish-white, or grey. The grain may be straight or interlocked, and the texture is coarse. It weighs about 580 kg/m$^3$ dried; moderately durable, it is susceptible to staining.

*A. adianthifolia.* Common in the secondary forest, the bole is often indented and twisted. The heartwood is light gold or light brown in colour, often with a greenish tinge. The sapwood, like *A. zygia* is very broad and of a creamy-white colour. The wood averages 580 kg/m$^3$ when dry, possesses a straight or interlocked grain and a moderately coarse texture. It is considered to be moderately durable.

*A. grandibracteata*, together with *A. zygia* is usually marketed in East Africa as red or white nongo, the distinction being one of colour.

13

Albizia is in abundant supply especially in Nigeria, but lack of demand has so far not resulted in proper commercial production. There would seem to be a case for segregation of the species on a weight basis ie, heavy albizia—640 kg/m$^3$ or more, and light albizia—under 640 kg/m$^3$. Uses could include construction, joinery and general carpentry.

## ALSTONIA

*Alstonia congensis* Engl .         Family : Apocynaceae
& *Alstonia boonei* De Wild.

**Other names**
patternwood, stoolwood (E and W Africa) ; mujua (Uganda) ; ahun, awun, duku (Nigeria) ; tsongutti (Zaire) ; sindru (Ghana) ; emien (Ivory Coast).

**Distribution**
Abundant throughout the humid forests in the Cameroons ; also found in Sierra Leone, Ivory Coast, Ghana, Nigeria and Zaire. It also grows in Central Africa and Uganda.

**The tree**
A tall tree, 30m or more in height and up to 1.0m in diameter, with a straight stem.

**The timber**
Nearly white when freshly cut, the timber darkens slightly on exposure. The sapwood which is not differentiated from the heartwood is very wide and up to 200mm ; soft, and light in weight when dried, the wood weighs about 400 kg/m$^3$.
The grain is generally straight, and the texture is fine, but the appearance of the wood is often marred by latex canals (slit-like holes about 6mm across) which often occur at regular intervals. The wood also is liable to staining.

**Drying**
Alstonia dries rapidly and well with practically no distortion.

**Strength**
Similar to obeche except in shock loading, when it is inferior.

14

## Working qualities
Works easily with all hand and machine tools, but because of the softness of the wood, sharp cutting edges are essential. Can be glued, stained and polished satisfactorily.

## Uses
Boxes, crates, rough carpentry for interior work.

# ANINGERIA

*Aningeria* spp.                    Family: Sapotaceae

## Other names
agnegre (Ivory Coast); landosan (Nigeria); mukali, kali (Angola); mukangu, muna (Kenya); osan (Uganda).
Note: veneer marketed under the misleading names of Tanganyika walnut and noyer de Bassam is believed to be aningeria.

Four species of *Aningeria* occur in tropical Africa, as follows, *A. robusta* is found in West Africa, *A. altissima* occurs in both west and east Africa, *A. adolfi-friederici* is widely distributed throughout East Africa and *A. pseudo-racemosa* also occurs in East Africa, principally in Tanzania.

## The tree
The trees are generally tall, commonly 30m to 36m in height, with straight, cylindrical boles some 1.2m to 2.4m in diameter, depending on the species and growth conditions. Bole lengths vary from 27m to 30m above the buttresses, but in *A. altissima* the boles may be shorter owing to the symmetrical winged buttresses which are often tall.

## The timber
The wood of the various species is somewhat similar in appearance, and not unlike birch. There is no clear distinction between sapwood and heartwood, except where sap-stain has developed, and the heartwood varies in colour from whitish, to a pale shade of brown, often with a pink tint. The wood is fairly plain in appearance, although quarter-sawn surfaces sometimes show a growth-ring figure, and where wavy grain is present, there is sometimes a slight mottle figure. The wood is

15

lustrous, sometimes with a faint cedar-like scent, and the various species are generally siliceous. The weight when dry varies from 510 kg/m³ to 570 kg/m³. The grain varies from straight to wavy, and the texture from medium to coarse.

## Drying
Dries easily and well both in the open and in the kiln apart from the tendency to blue-stain in the early stages of air drying.

## Strength
About 20-25 per cent weaker than birch in bending and compression, but comparable in hardness on side grain.

## Durability
Perishable.

## Working qualities
Reports on the sawing properties of the timber are contradictory, varying from hard due to silica content to easy, but a moderate to severe abrasive action on tools and cutters should be anticipated. In cross-cutting and boring, adequate support is needed to prevent chipping out, and care is needed in planing in order to obtain a smooth finish. The wood takes and holds nails and screws well, and can be glued, stained and polished. The wood can be peeled or sliced for veneer, but for thick veneer for plywood, a softening treatment at 85°C is necessary and encourages a high yield of good veneer which can be dried without distortion and splitting.

## Uses
Plywood and veneer, joinery, and general interior utility.

## ANTIARIS

*Antiaris toxicaria* Lesch.                     Family : Moraceae
Syn. *A. welwitschii* Engl.

## Other names
oro, ogiovu (Nigeria) ; chenchen, kyenkyen (Ghana) ; kirundu (Uganda) ; diolosso (Cameroons) ; ako (Ivory Coast).

## Distribution
Widely distributed in the high forest zone of West, Central and East Africa.

## The tree
Antiaris may reach a height of 36m to 45m with a diameter of 0.6m to 1.5m. Clear boles up to 21m are common. There are usually no buttresses.

## The timber
The timber is white to yellowish-grey, with little difference between heartwood and sapwood; soft, light in weight, averaging about 430 kg/m$^3$ when dried, and of a fibrous nature. Not very durable nor strong. The timber should be converted and dried as rapidly as possible to prevent staining.

## Working qualities
A timber which is easy to saw and work, but liable to tear and pick up on quartered surfaces unless the cutting angle is reduced to 20°. It has good nailing, staining and gluing characteristics.

## Uses
It is used for light construction, and as quarter-sliced veneer for furniture. It is not suitable for all-veneer plywood because it is too coarse and too soft.

# AVODIRÉ

*Turraeanthus africanus* Pell                    Family: Meliaceae
and *T. vignei* Hutch. and J. M. Dalz.

## Other names
apeya, appayia, wansenwa (Ghana); apaya (Nigeria); engan (Cameroons); lusamba, esu, songo (Zaire).

## Distribution
These species appear relatively restricted to a coastal belt stretching from Ghana westwards as far as the Bandama River on the Ivory Coast. It is a gregarious species and has also been recorded in Angola.

## The tree
A medium-sized tree, from 18m to 30m high with a diameter of about 0.6m. The habit of growth is not good, the trunk often being crooked and irregular.

## The timber
There is generally no clear distinction between heartwood and sapwood; the wood is pale yellow with a natural lustre, and darkens to pleasing golden-yellow colour. The grain is sometimes straight but often wavy or irregularly interlocked, producing a beautiful mottled figure when quarter sawn. The appearance of the figured wood suggests East Indian satinwood except it has a more open texture. The weight is about 560 kg/m³ when dried, ie, about the same as African mahogany.

## Drying
Avodiré can be dried fairly rapidly, but has some tendency to distort, and existing shakes to extend.

## Strength
It is a strong, tough and elastic timber in proportion to its weight.

## Durability
Non-durable.

## Working qualities
It works fairly easily by both hand and machine tools. In planing it tends to pick up owing to the interlocked grain, and a cutting angle of 20° or less is desirable for a smooth finish. The timber has fairly good nailing, screwing, and gluing properties, takes stain well, and gives good results with the usual finishing treatments.

## Uses
Figured material is satisfactory for sliced veneer, high-grade cabinet work and panelling. Plain stock is useful for plywood manufacture, but selection of suitable logs is not always easy.

# AYAN

*Distemonanthus benthamianus* Baill.     Family: Leguminosae

## Other Names
anyaran, anyanran, anyan (Nigeria); barré (Ivory Coast); eyen (Cameroons); oguéminia (Gabon); movingui (Benin).

## Distribution
Ayan is widely distributed in West Africa from the Ivory Coast to Gabon and Zaire.

## The tree
This is a slender tree attaining a height of more than 30m but seldom more than 0.75m diameter. The bole is straight, cylindrical and free from buttresses.

## The timber
The pale yellow sapwood is narrow and not clearly defined; the heartwood is pale to bright yellow or yellow-brown. It is fine textured, the grain is often interlocked, sometimes wavy, and the wood has a very lustrous surface. Moderately hard and heavy it weighs about 690kg/m³ when dried.

## Drying
Dries fairly well with little tendency to split and warp.

## Durability
Moderately durable.

## Strength
Good strength properties, especially in compression along the grain and in bending.

## Working qualities
Works well with hand and machine tools; there is a tendency to pick up when planing quarter-sawn stock owing to inter-locked grain. It takes nails and screws fairly well, finishes excellently, and takes a high polish if sufficient filler is used.

## Uses
Furniture, cabinet work, ships fittings, flooring, interior joinery. Owing to the presence of a yellow extractive, which in moist conditions can produce a yellow dye, the wood is not suitable for use as clothes airers, laundry equipment.

# BANGA WANGA

*Amblygonocarpus andongensis*             Family : Leguminosae
Exell & Torre

## Other names
None.

## Distribution
The natural distribution of *Amblygonocarpus* is restricted to the savannah area of Uganda and adjacent territories, extending southwards to Mozambique.

## The tree
The tree is small to medium, occasionally reaching 20m in height, with a diameter of about 0.5m.

## The timber
The narrow, greyish-white sapwood is clearly defined from the heartwood which is warm brown or brownish-red in colour, with a subdued 'partridge wing' figure on tangential surfaces due to soft tissue which surrounds the pores and links several together.
The wood is hard and heavy, weighing about 1040kg/m³ when dried. The grain is rather irregular and sometimes interlocked, but the texture is fine.

## Drying
Dries slowly owing to its hardness, and requires extreme care if checking and distortion are to be avoided.

## Strength
A heavy, hard, and strong timber, with a high resistance to impact and abrasion.

## Durability
Probably very durable.

## Working qualities
When dry, it is reputed to work reasonably well, but requires a reduced cutting angle in planing and moulding. There is a tendency to 'ride' on cutters.

## Uses
At present, it is best known for heavy duty flooring, but is used for sleepers in East Africa.

## BERLINIA

*Berlinia* spp.,                                    Family : Leguminosae
including *B. bracteosa* Benth.,
*B. confusa* Hoyle.,
and *B. grandiflora* Hutch. & Dalz.

## Other names
ebiara (Liberia and Gabon) ; ekpogoi (Nigeria) ; abem (Cameroons).

## Distribution
The various species of *Berlinia* grow throughout West Africa in various types of forests. The most important timber species occur in the high forest belt of Sierra Leone, Liberia, the Ivory Coast, Ghana, Nigeria and the Cameroons.

## The tree
The trees are from 24m to 42m high, and may be up to 1.2m in diameter, with an average of 0.75m at maturity. The trunk is usually without buttresses, but it may be fluted at the base. The bole is reasonably straight, clear and cylindrical.

## The timber
The heartwood varies from light red to dark red-brown, with dark purple or brown irregular streaks. The sapwood is pink when freshly cut, but turns whitish or greyish on exposure. Traumatic vertical gum ducts are frequently present, appearing on the cross section in arcs of varying length. The texture of the wood is coarse, and the grain is usually interlocked and sometimes very irregular. Brittleheart may occur in large logs.
The timber is moderately hard and moderately heavy ; the weight is variable but averages 720 kg/m$^3$ when dried.

## Drying
Berlinia dries slowly and well and, except for the occasional piece, without distortion. Discoloration from mould growths which tend to develop during kiln drying is a frequent problem.

## Strength
Its strength properties compare favourably with English oak.

## Durability
Moderately durable.

## Working qualities
Since the sapwood band in berlinia is often very wide (100mm to 150mm is common) the amount of sapwood present in the parcel will have a bearing on the working qualities. Sapwood that has relatively straight grain and the less dense heartwood can be worked fairly easily with both hand and machine tools and have a moderate blunting effect on their cutting edges. The tearing that occurs in planing and moulding can be reduced by reducing the cutting angle to 20° unless the wood has wavy grain, when a smooth finish is not possible. When the denser heartwood is cut there is a tendency for ripsaws to vibrate in the cut. Most other operations are satisfactory. Sapwood with highly irregular grain is difficult to stain and polish, but other material stains well. It glues satisfactorily and nails fairly well.

## Uses
Carpentry and general cabinet-making. The timber is suitable for the same uses as oak, except for bending. It is also used in vehicle bodies.


# AFRICAN BLACKWOOD

*Dalbergia melanoxylon*        Family: Leguminosae
Guill. and Perr.

## Other names
mpingo (Tanzania).
Although this timber is sometimes given the name 'African ebony', it is misleading as the true ebonies have the generic name *Diospyros*; it would therefore be more correct to call it a rosewood since it is a species of *Dalbergia*. The standard name is African blackwood.

## The tree
This is a small tree growing to about 9m high, with a clear bole

rarely over 2.5m high and a diameter of about 200mm rarely exceeding 300mm. The tree is occasionally multi-stemmed.

## The timber
The sapwood is narrow, white in colour, and clearly defined from the dark heartwood, which is dark brown with predominant black streaks which give an almost black appearance to the wood. It is straight grained and extremely fine textured, hard and more dense than rosewoods generally, weighing about 1200 kg/m$^3$ when dried.

## Drying
It is generally partially dried in log or billet form and then converted and stacked under cover to complete the drying. The timber dries extremely slowly and heart shakes are very common. In general the wood needs to be carefully handled to minimize checking.

## Durability
Very durable.

## Working qualities
In spite of its hardness it works quite easily, and takes an excellent finish. It is however, moderately hard to saw, and requires drilling for nails and screws.

## Uses
Ornamental turnery, chessmen, carved figures, walking sticks, inlay work, brushbacks, knife handles, and pulley blocks. Its oiliness and resistance to climatic changes commend it for woodwind instruments in preference to ebony, and it is used in the manufacture of bagpipes, clarinets, piccolos and flutes.

## BOXWOOD

*Buxus macowani* Oliv.　　　　　　　　Family: Buxaceae
and *Gonioma kamassi* E. May.　　　　Family: Apocynaceae

The trade name boxwood, originally the English equivalent of *Buxus sempervirens*, has been extended to cover a number of botanically unrelated species with wood resembling true boxwood in general character, eg *Gonioma kamassi, Gossypiospermum praecox* and *Phyllostylon brasiliense.*

## Other names
East London boxwood, Cape box (S Africa), *Buxus macowani.* Knysna boxwood, kamassi boxwood, (S Africa) *Gonioma kamassi.*

## Distribution
East London boxwood occurs in a strip of forest along the south-east coast of Cape Province, South Africa, while Knysna boxwood occurs in the coastal forests on the south coast of Cape Province.

## The tree
East London boxwood is a small tree with a clean bole of about 4.5m to 6m and a diameter of about 150mm at breast height when mature.
Knysna boxwood is a slightly larger tree attaining a height of 6m and a diameter of 300mm.

## The timber
The wood of both species is very similar to true boxwood (*Buxus sempervirens*) and while *G. kamassi* weighs about 880 kg/m³ *B. macowani* weighs about 960 kg/m³ when dried.

## Drying
The main defect encountered in drying is the development of very fine surface checks, which however, tend to penetrate deeply into the wood, and may open as drying proceeds. Drying is best carried out under cover; small logs being left in the round, and larger ones either cut in halves, or converted to dimension stock. The timber should be allowed to dry very slowly; reports from South Africa suggest four to five months in which to dry 25mm stock in the open air.

## Strength properties
No strength data are available, but are probably similar to European boxwood.

## Durability
Probably durable.

## Working qualities
Fairly hard to work, especially with hand tools, but dulling effect is only moderate. Can be machined to a fine finish, but a reduction of cutting angle to 20° with an increased loading on pressure bars and shoes will minimize the tendency for the wood to 'ride' on the cutters during planing and moulding. Can be turned well; should be pre-bored for nailing and screwing; glues, stains, and polishes well.

## Uses
Generally the same as for European boxwood; turnery, engraver's work, small rollers and shuttles for the textile industry.

# BUBINGA

*Guibourtia* spp.                    Family: Leguminosae
principally *G. demeusei* J. Léon,
but including *G. pellegriniana* J. Léon,
and *G. tessmannii* J. Léon

## Other names
kévazingo (Gabon); essingang (Cameroons).

## Distribution
Mostly found in the Cameroons and Gabon.

## The tree
The trees attain a height of 24m to 30m with a diameter of 1.0m or more.

## The timber
The heartwood is light red-brown attractively veined with pink or red stripes; the sapwood is lighter in colour. The wood is hard and heavy, weighing from 800 to 960 kg/m³ when dried. It has a fine texture.

## Uses
Almost exclusively used for decorative veneer; when sliced it is

called bubinga, and when rotary cut (which gives a somewhat different figure) it is called kévazingo.

## 'EAST AFRICAN CAMPHORWOOD'

*Ocotea usambarensis* Engl.                    Family : Lauraceae

**Other names**
camphor, muzaiti, muura, munganga, mutunguru.

**Distribution**
Kenya and Tanzania.
Although the timber has a distinct scent of camphor, it should not be confused with Borneo or Sabah camphorwood (*Dryobalanops* spp), or true camphorwood (*Cinnamomium camphora*).

**The tree**
The largest tree of Kenya, attaining a height of 36m and a diameter of 1.5m to 2.0m and sometimes as great as 3.0m. It is a tree of the mountain and rain forests, occurring on the southern and eastern slopes of Mount Kenya and in the Aberdare Range. In Tanzania it is found on Mount Kilimanjaro, Usambara and Upare.

**The timber**
The timber, when fresh, is yellowish with a greenish or brownish tinge, darkening to deep brown on exposure. The sapwood is paler, but not always clearly defined. It generally lacks figure, except when quarter cut when it has a pronounced stripe owing to interlocking grain. Texture moderately fine ; moderately hard, and moderately heavy, weighing about 610 kg/m³ when dried. The timber has a camphor like odour when fresh, but this disappears in due course.

**Drying**
Needs care in air drying as it is liable to warp and twist if this process is too rapid. Kiln drying can be carried out slowly, and with little degrade, but there may be some difficulty in removing the moisture from the centre of thick stock, especially when quarter sawn.

26

## Strength
For its weight, East African camphorwood has extremely good strength properties being superior to American mahogany in most categories.

## Durability
Very durable. Reported to be termite resistant.

## Working qualities
The timber is easy to work and saw, but needs care in planing quarter-sawn surfaces to prevent the interlocked grain picking-up; a cutting angle not greater than 20° should be used.
The timber can be successfully peeled; it takes nails and screws well, and glues satisfactorily.

## Uses
Furniture, panelling, interior and exterior joinery, flooring, light constructional purposes.

# CANARIUM, AFRICAN

*Canarium schweinfurthii* Engle.          Family: Burseraceae

## Other names
papo, elemi (Nigeria); abeul (Gabon); abel (Cameroons); mupafu, mbidinkala, mwafu (Uganda).

## Distribution
It has a wide distribution stretching from Sierra Leone to Angola and Uganda.

## The tree
Canarium attains a large size, sometimes exceeding 30m in height, and 1.2m in diameter above the thickened base. The average length of the clear bole is often between 14m and 15m but may reach 23m in good specimens.

## The timber
Light brown or pinkish brown, resembling gaboon (okoumé) in general appearance. The surface is lustrous and the wood is distinctly scented when freshly sawn. The texture is somewhat

coarse and may be woolly; the grain is often spiral. Large logs are liable to contain brittleheart. It weighs on average 530 kg/m³ when dry, which is slightly heavier than gaboon.

## Drying
Not unduly difficult to dry although some distortion and even collapse may occur.

## Durability
Non-durable.

## Strength
Strength properties are similar to agba, especially in regard to stiffness, resistance to compression and shear strength, but it is slightly weaker in bending and shock resistance.

## Working qualities
Canarium is a difficult timber to saw; owing to the presence of silica in the wood it blunts saw teeth extremely rapidly.
Planing and moulding is comparatively easy, but sharp cutting edges must be maintained, and a reduced cutting angle of 20° normally provides clean smooth surfaces.

## Uses
Canarium is a utility wood for interior joinery and carpentry. It produces good veneer but usually of core quality suitable for plywood.

# CEIBA

*Ceiba pentandra* Gaertn.                    Family: Bombacaceae.

## Other names
silk cotton tree, okhar (Nigeria); kapokier, fuma (Zaire), fromager (France).

## Distribution
Widely distributed in the Eastern hemisphere tropics, it occurs particularly in Nigeria, where it springs up readily in clearings.

## The tree
Ceiba grows to a height of 30m to 36m and diameters of 1.2m to 2.0m are common. The trunk is heavily buttressed.

## The timber
The wood is whitish with yellowish streaks, or greyish, often with a pinkish tinge. It is often cross-grained and the texture is coarse. Soft, and light in weight, it weighs on average 350 kg/m$^3$ when dried. It is a dull looking wood, very absorbent, and prone to discoloration unless rapidly extracted, converted and dried.

## Drying
Dries well with little degrade such as distortion, splitting and checking.

## Durability
Non-durable.

## Working qualities
Easily worked, but rather difficult to obtain a smooth finish.

## Uses
Sound insulation eg cabin panelling, core stock, and simple carpentry and joinery.

Note : A closely related genus produces West African bombax. This is principally the product of *Bombax buonopozense* P. Beauv. In its general characteristics, colour, and weight, it is very similar to ceiba. Two other species ie *B. breviscuspe* Sprague, known as kondrotti and *B. chevalieri* Pellegr., known as alone have slightly heavier timber, are medium brown in colour and are occasionally shipped. The uses for bombax are the same as for ceiba.

## CELTIS, AFRICAN

*Celtis* spp.                                        Family : Ulmaceae
principally *C. adolphi-friederici* Engl.,
*C. milbraedii* Engl.,
*C. zenkeri* Engl.
syn *C. soyauxii* Engl.

## Other names
esa (Ghana) ; ita, ohia, ba (Nigeria) ; shiunza, chia, mudengwa, kiambo (Uganda) ; kerrua, chepkelelet (Tanzania).

## Distribution
Celtis is very common in parts of Nigeria and Ghana, and extends from West Africa through Central Africa to Uganda, Tanzania, and part of Kenya.

## The tree
An evergreen tree with a straight, fairly heavily buttressed bole. Height 30m and diameter of 1.0m or occasionally more.

## The timber
The timber is whitish or greyish in colour turning yellowish or straw coloured on exposure. The grain is irregular, but is sometimes straight, and the texture is fairly fine. The weight averages 800 kg/m³ in the dry condition.

## Drying
Needs care in drying, as it is apt to warp and split.

## Strength
For its weight, the strength properties of celtis are, on the whole, rather above the average. In many of its properties it compares favourably with ash.

## Durability
Non-durable, and is rapidly attacked by staining fungi in adverse circumstances.

## Working qualities
Works well with moderate ease in machine operations, but is slightly hard to work by hand. Straight grained material finishes cleanly in general but there is some tendency to tear out in planing when interlocked grain is present. A cutting angle of 15° is recommended. The timber is inclined to split in nailing, but takes glue, stain and polish well.

## Uses
Celtis has been used successfully as a substitute for ash. It is a general utility timber suitable for interior joinery and carpentry. It is a very good flooring material, having a high resistance to wear; there is little surface breakdown, the surfaces wearing smooth, and could be a substitute for maple flooring.

# CORDIA

*Cordia* spp. Family: Boraginaceae
including *C. abyssinica* R.Br.,
*C. millenii* Bak.,
and *C. platythyrsa* Bak.

## Other names
mukumari, mugona, mringaringa, mungoma (Kenya and Tanzania) ; omo (Nigeria).

## Distribution
The African species of *Cordia* occur as a deciduous tree of semi-tropical rain forests in Kenya and Tanzania, and to a lesser extent in West Africa, generally in Nigeria.

## The tree
A medium-sized tree averaging 10m high and 0.6m diameter. The bole is generally irregularly shaped.

## The timber
The wood is variable in colour, ranging from rich golden-brown to fawn with dark streaks, usually darkening on exposure to a light reddish brown not unlike the associated freijo (*C. goeldiana*) which grows in Brazil. The sapwood is cream coloured. The medium sized rays in cordia promote a mottle figure on quarter-sawn surfaces. A medium-textured wood, generally fairly straight-grained, but occasionally interlocked. Light and moderately soft, it weighs about 480 kg/m$^3$ when dried.

## Drying
Dries easily and well without undue splitting and warping.

## Durability
Moderately durable.

## Working qualities
The timber is easy to work but sharp tools are needed to prevent the surface from becoming woolly. Nails well, and can be polished effectively if care is taken.

## Uses

Furniture, cabinets and library fittings. In Africa it is often used for making traditional drums because of its resonant quality, and for generations has been a canoe making wood in Nigeria. Could also be considered useful for general joinery purposes.

## *DACRYODES* SPECIES

Family : Burseraceae

The following species occur in West Africa :
*Dacryodes buettneri* ; known as ozigo (Gabon)
*D. edulis* ; known as ollem
*D. igaganga* ; known as igaganga (Gabon)
*D. klaineana* ; known as adjouaba
*D. le-testui* ; known as mouvendo
*D. normandii* ; known as ossabel (Gabon)
*D. pubescens* syn *D. heterotrichia* ; known as safoukala (Congo)

**The tree**
The trees vary in height according to species, from 24m to 36m with diameters ranging from 0.3m to 1.5m.
The boles are generally straight and cylindrical, with the odd specimen, particularly in *D. klaineana*, being irregular. There are generally no buttresses, but some boles may be fluted.
**The timber**
*Dacryodes* is botanically associated with gaboon (*Aucoumea klaineana*), and there is some similarity in the general appearance of the wood of all the species.
**Ozigo:** sapwood not clearly demarcated from the heartwood which is grey-buff in colour. The grain may be interlocked or crossed, or straight, and the texture is rather coarse. The wood has a slight lustre and a ribbon figure on quarter-sawn surfaces. Weight about 650 kg/m³ when dry.

**Ollem:** similar to ozigo in appearance and colour, but with a medium texture, and with some included resin. Weight about 650 kg/m³ when dry.

**Igaganga:** sapwood 25–37mm wide, lighter in colour than the heartwood which is a pinkish-buff colour. The grain is usually interlocked, seldom straight, and the texture varies from medium to fine. The wood sometimes produces a decorative figure, but it contains silica and gum ducts. Weight about 650 kg/m$^3$ when dry.

**Adjouaba:** sapwood paler than the heartwood which is a grey or pink colour. The grain is straight to interlocked, and the texture fine. The wood contains resin. Weight 810 kg/m$^3$ when dry.

**Mouvendo:** sapwood barely distinguishable from the pink-buff or grey-buff heartwood. Grain wavy or straight, texture medium. Weight about 650 kg/m$^3$ when dry.

**Ossabel:** sapwood not clearly demarcated from the heartwood which is a pink-buff colour with an occasional grey tinge. The grain is often interlocked, and the texture medium to coarse. Weight about 650 kg/m$^3$ when dry.

**Safoukala:** sapwood very wide, often extending over half the diameter of the log. The general colour may be pale pink, yellow, or grey with the heartwood a little darker. The grain may be straight or interlocked, and the wood often shows attractive ribbon, or small stripe figure. The texture is moderately fine, and the wood weighs about 750 kg/m$^3$ when dry.

## Drying
All species dry quite well, but care is needed in air drying to avoid surface checking.

## Strength
Similar to gaboon, but more fissile.

## Durability
Non-durable.

## Working qualities
All species work and machine reasonably well, although some dulling of cutting edges can be expected due to the silica content. There is a tendency for the grain to pick up in planing and moulding, and a reduction of cutting angle to 15–20° is usually essential, but the wood sands to a good finish. All

species peel well for veneer, and plywood manufactured in Ghana from adjouaba has been reported to compare favourably with Canadian yellow birch. The timber has good nailing and screwing properties, and can be glued, stained and polished satisfactorily.

## Uses
Interior joinery, flooring, plywood, boxes and crates.

# DAHOMA

*Piptadeniastrum africanum*        Family : Leguminosae
(Hook f.) Brenan.
Syn. *Piptadenia africana* Hook, f.

## Other Names
ekhimi, agboin (Nigeria); dabéma (Ivory Coast); toum (Gabon); atui, tom, bokungu (Cameroons); banzu, musese, singa (Zaire); mpewere, mapewere (Uganda); mkufi (Tanzania).

## Distribution
Dahoma is a common species in the rain forests of West, Central, and parts of East Africa, but is more prevalent, and attains its best development in the mixed deciduous forest.

## The tree
Dahoma is a large tree usually 36m or more high, but the large, spreading, plank buttresses reduce the usable bole to something like 10m to 15m with a diameter of 1.0m or slightly more. The bole is usually straight and cylindrical.

## The timber
Sapwood is whitish to greyish-red, sharply defined from the heartwood which is light to golden-brown, very similar in appearance to iroko. A ribbon grain produces zones of light and dark colour. When freshly cut there is an unpleasant smell resembling ammonia and the sawdust is sometimes irritating to the eyes.
The texture is coarse and woolly, but uniform. It weighs from 560 to 710 kg/m$^3$ when dried.

## Drying
Rather difficult to dry, but variable in this respect. Dries slowly, and there is a tendency to distort, and a liability to collapse in some thick material.

## Strength
The strength values are good, resembling those of iroko, but because of interlocking grain it is not suitable for use in small sections where strength in bending is a prime requirement.

## Durability
Moderately durable; reported to be resistant to termites.

## Working qualities
Works fairly well but with appreciable dulling effect on tools. There is a tendency to pick up in planing, and a cutting angle of 10° is recommended. Takes a good finish and may be screwed and nailed, with only a slight tendency to split when nailed near the edges.

## Uses
Domestic flooring, vehicle bodies and floors, structural work and marine uses.

# DANTA

*Nesogordonia papaverifera* Capuron,               Family: Tiliaceae
Syn. *Cistanthera papaverifera* A. Chev.

## Other names
otutu (Nigeria); kotibé (Ivory Coast); olborbora (Gabon); ovoué (Cameroons); tsanya (Zaire).

## Distribution
Danta grows in the mixed deciduous forests in southern Nigeria, the Ivory Coast and Ghana.

## The tree
A fairly large tree some 27m to 30m high with a long, cylindrical bole, about 15m high, above the sharp buttress, and with a diameter of about 0.75m.

35

## The timber
The sapwood is light brown with a pinkish tinge and is sharply defined from the heartwood, which is reddish-brown and has a lustrous surface; it has a fine texture, and the grain is typically interlocked and this produces a ribbon figure, the striped appearance on quarter-sawn material resembling sapele. It is hard and fairly heavy, being about 750 kg/m³ when dried. Planed surfaces have a somewhat greasy feel.

## Drying
Dries fairly well with some tendency to warp. Care must be taken to avoid over-rapid drying or there may be a danger of casehardening with 'ribbing' of the surface.

## Strength
It is a very strong and elastic timber and some of its mechanical properties are similar to those of European ash. It is however, significantly weaker than ash in its resistance to impact loads and cannot be recommended in place of ash for the most exacting purposes.

## Durability
Moderately durable.

## Working qualities
Danta works well with most tools, but has a tendency to pick up on quarter-sawn material. It turns excellently and takes a good finish and polishes well.

## Uses
Lorry bodies, coach and wagon work, general construction and for purposes where a strong, durable wood is required. It wears smoothly and has a good resistance to abrasion, and is therefore suitable for flooring. It is used for telegraph cross-arms and railway sleepers. In Nigeria it ranks second to celtis in its uses for tool handles, eg files, screwdrivers, etc, and is considered to be superior to mansonia and sycamore as an etching timber in graphic art.

# DIFOU

*Morus mesozygia* Stapf.                    Family : Moraceae

## Distribution
Tropical Africa, particularly the Ivory Coast.

## The tree
Grows to a height of 27m to 36m and a diameter of 0.6m to 0.9m.
The tree has wide spreading root ridges and produces a straight cylindrical bole some 18m in length.

## The timber
The sapwood, which is wide in young trees, is pale grey or white in colour, and the heartwood is yellow when freshly cut, darkening with age to coffee brown. It often shows a mottle figure. The grain is shallowly interlocked, and the texture is fine. The wood is often confused with the related iroko, which it resembles in appearance, but difou is a slightly heavier wood, and the texture is finer than that of iroko which also has rather larger pores. Difou weighs about 760 kg/m³ when dry.

## Drying
Dries quite well and rapidly, without undue distortion.

## Strength
The strength properties are superior to those of iroko, and more nearly approximate those of okan.

## Durability
Moderately durable.

## Working qualities
Although hard and heavy, difou saws quite satisfactorily, although cutting edges generally tend to dull fairly rapidly. In planing and moulding a reduction of cutting angle to 15° is essential to avoid grain tearing, especially on quarter-sawn surfaces. The wood glues and polishes satisfactorily, and takes and holds nails and screws well. It produces good veneer.

## Uses
Heavy construction, flooring, vehicle bodies, furniture, sporting goods, agricultural implements, veneer and plywood, joinery and turnery.

# EBONY, AFRICAN

*Diospyrus* spp.                                Family: Ebenaceae
including *D. crassiflora* Hiern.,
and *D. piscatoria* Gurke.

## Other names
Cameroons, Gabon, Kribi, Madagascar, Nigerian ebony, according to origin.

## Distribution
Limited range in southern Nigeria, Ghana, Cameroons and Zaire.

## The tree
A small to medium-sized tree attaining a height of about 18m and a diameter of about 0.6m.

## The timber
One of the best jet-black varieties is believed to be *D. crassiflora* It is a very heavy timber weighing 1030 kg/m$^3$ when dried. Other species produce handsome black and brown striped varieties.

## Drying
In general African ebony air dries quite readily, but there is a tendency for surface checks to develop. In small sizes, the timber kiln dries fairly quickly and well, with little tendency to split or distort.

## Durability
Probably very durable.

## Working qualities
The black heartwood is inclined to be brittle and is rather hard to work, and has a considerable dulling effect on cutters which should have their cutting angle reduced to 20° to avoid picking up when curly grain is present. It takes a fine polish.

## Uses
Turnery, inlaid work, fancy articles, brush backs.

# EKABA

*Tetraberlinia bifoliolata*                    Family : Leguminosae

## Other name
ekop ribi.

## Distribution
Found in mixed forests, often in small clusters, particularly in the Cameroons and Gabon.

## The tree
A tall tree, it reaches a height of 45–50m and a diameter varying from 0.9m to 2.0m.
The bole is generally straight, clear and cylindrical, and the base is sometimes swollen or fluted.

## The timber
The sapwood varies in width from 25mm to 125mm some logs having wide sapwood and brittle heart. There is not much difference by colour between sapwood and heartwood, the wood generally being a pale buff when first cut, turning rapidly to pink, with darker streaks. There is often a greyish tinge to the wood, and hard black kino is often present. The grain is usually interlocked or irregular, and the texture is medium to rather coarse. The wood weighs 720 kg/m$^3$ when dry. Wet wood tends to stain when in contact with iron.

## Drying
The timber dries rather slowly but well, with only isolated instances of moderate distortion.

## Strength
No data are available.

## Durability
Moderately durable.

## Working qualities
Works and machines with moderate ease, but there is a tendency for grain tearing to occur in planing and moulding, and a reduction of cutting angle to 20° is necessary in order to obtain

a clean finish. The wood takes stains and polish well, and gluing is satisfactory, although there may be  tendency for the glue to show on the wood as dark blotches.

## Uses
Flooring, veneer, furniture, cabinets, joinery and turnery.

# EKEBERGIA

*Ekebergia rueppeliana*                     Family : Meliaceae
Fresen. ex A. Rich.

## Other names
monko, mfuari, mbo, teldet, mukongu, mununga.

## Distribution
Found in Tanzania, Uganda, Kenya and southern Rhodesia, commonly in the rain forests.

## The tree
A large tree up to 30m high, with a clear bole of 10m and a diameter of 1.0m. The trunk is often fluted and crooked.

## The timber
The wood is generally a light pinkish-brown with irregular darker lines. The grain is straight, and the texture is medium to coarse. It weighs about 545 kg/m³ when dried.

## Drying
It is reported to dry fairly well, without undue warping and splitting. It is liable to be attacked by staining fungi if not con-verted and dried rapidly.

## Strength
No information available.

## Durability
Non-durable.

## Working qualities
The timber is easy to work, nails well, and is reported to take a good finish.

## Uses
Interior joinery and furniture.

# EKKI

*Lophira alata* Banks ex Gaertn.          Family: Ochnaceae

## Other names
kaku (Ghana); azobé (Ivory Coast); bongossi (Cameroons);
akoura (Gabon); eba (Nigeria); hendui (Sierra Leone).

## Distribution
Grows in West Africa from Sierra Leone to Nigeria and the
Cameroons. It is a tree of the heavy rain forests and swamps.

## The tree
Ekki is a large tree, attaining a height of 45 to 50m and a
diameter of 1.5m.

## The timber
The sapwood is pale pink and sharply defined from the heartwood, which is red-brown to dark brown with a somewhat
speckled appearance due to white deposits in the pores. The
grain is usually interlocked and the texture is coarse. The wood
is extremely hard and heavy, weighing 960 to 1120 kg/m$^3$
when dried.

## Drying
Very difficult to dry and generally shakes badly; serious degrade
is likely, especially surface checking and end splitting. Needs to
be piled with special care.

## Durability
Very resistant to decay; one of the most durable woods yet
known in West Africa.

## Working qualities
Very difficult to work with hand tools, but can be worked by
machines with less trouble. Must be pre-bored for nailing.

## Uses
Too hard for some purposes, but is suitable for heavy construc-

tion especially wharves, bridge building and decking, sleepers, flooring—especially heavy duty flooring. It is ideal for all forms of marine work for piling, sea defences, groynes and jetties, and any use where high strength and durability is a prime requirement.

# EKOUNE

*Coelocaryon klainei*　　　　　　　　　Family : Myrtaceae
Syn *C. preussii*

## Distribution
West Africa principally in Gabon and Congo.

## The tree
Grows to a height of 30m and a diameter of 0.3m to 0.9m with a clean, straight, and cylindrical bole, usually unbuttressed, of some 12–18m in length.

## The timber
There is generally little difference in colour between sapwood and heartwood ; the sapwood is very wide, white or pale pinkish-yellow in colour, darkening on exposure. The heartwood is pale brown, occasionally reddish-yellow, with dark markings. The wood is lustrous, with a straight grain, and with a medium to fine texture. It varies in weight between 650 kg/m³ to 720 kg/m³ when dry.

## Drying
Dries easily and well, but is prone to sap stain and should be anti-stain dipped.

## Strength
Similar to African mahogany in most strength categories.

## Durability
Non-durable.

## Working qualities
Saws easily, and works well with all tools. Takes nails, screws, glue, and stains and polishes satisfactorily. It is a good peeling timber.

## Uses
Furniture, veneer, plywood, interior trim and joinery, turnery.

# ERIMADO

*Ricinodendron heudelotii*                    Family : Euphorbiaceae
Pierre ex Pax.
Syn. *R. africanum* Muell. Arg.

## Other names
wama (Ghana) ; sanga sanga (Zaire) ; essessang (Cameroons, Gabon) ; mungenge (Angola).

## Distribution
*Ricinodendron* is a tree typical of the secondary forest, and it is quite common on the site of abandoned farms. It is widely dispersed in tropical Africa from Guinea to Angola, Zaire, Uganda, and other areas of East Africa.

## The tree
It is considered to be one of the fastest growing African trees, variable in size ; it may become very large, up to 30m or more tall, and almost 1.5m in diameter, but sometimes it is a small tree only 6m to 10m high. It is usually medium sized however, with a height of up to 21m and a diameter of about 0.75m. The trunk is cylindrical, sometimes with very short buttresses.

## The timber
The heartwood is not distinct from the sapwood. It is white or straw-coloured, but often appears greyish because of fungal staining. The texture is rather coarse, and the grain is straight. The wood is very soft and extremely light, weighing between 200 and 350 kg/m$^3$ when dried.

## Drying
No information, but it is reputed to have a high volumetric shrinkage, and not to remain stable when worked.

## Durability
Perishable.

## Working qualities
Easy to saw, but difficult to plane because of lifting of the fibres.

The assembly of parts is not firm; nails and screws enter easily, but they do not hold well. Unsuitable for painting because of the wood's absorbency.

## Uses
It is a good substitute for balsa, or cork, but because of its long, thin-walled fibres, is more suitable for paper pulp or cellulose extraction for rayon or fibre board.

# ESIA

*Combretodendron macrocarpum*          Family: Lecythidaceae
(P. Beauv.) Keay.
Syn. *C. africanum* Exell.

## Other names
owewe (Nigeria); minzu (Zaire); abale (Ivory Coast); abine (Gabon).

## Distribution
Grows throughout West Tropical Africa from Guinea to Zaire and Angola. It is infrequent in the dry high forests, but fairly common in the wet forest areas, particularly in Nigeria.

## The tree
A large tree up to 36m high and about 1.0m in diameter. The bole is straight and cylindrical. It is unbuttressed, but may be swollen at the base.

## The timber
The sapwood is yellowish-white, sharply defined from the heartwood, and about 75mm wide. The heartwood is rose coloured when freshly cut, but when exposed to the air it darkens to a reddish-brown. Darker veins or streaks give it a rather speckled appearance. The wood has a strong, unpleasant odour when freshly cut, which disappears on drying.
The wood is hard and heavy and weighs about 800 kg/m³ when dried. The grain is straight to interlocked, and the texture is moderately coarse.

## Drying
Difficult to dry without distortion. The wood dries slowly and is likely to check and split. End-splitting, surface checking, and shakes may be serious. It is improbable that this timber can be

kiln dried satisfactorily from the green condition. It is classified as having large movement values.

## Durability
Durable.

## Working qualities
Rather difficult to work; saw teeth tend to overheat due to fine sawdust adhering to the packing of circular saws. In planing, a cutting angle of 20° is necessary to prevent tearing of the surfaces. The wood stains and polishes well, but grain filling is usually necessary.

## Uses
Heavy or rough construction work, railway sleepers. The unfavourable characteristics of esia prevent its use for more exacting work.

# GABOON

*Aucoumea klaineana* **Pierre.**　　　　　**Family: Burseraceae**

## Other names
okoumé (Gabon); mofoumou, n'goumi (Equatorial Guinea).

## Distribution
The natural distribution of gaboon is fairly restricted, being found mainly in Equatorial Guinea, Gabon, and the Congo.

## The tree
The tree attains very large sizes and may be 1.0m to 2.5m in diameter at the base.

## The timber
Gaboon is one of the most useful hardwoods and is used in large quantities for plywood and blockboard. As it has a fairly close resemblance to African mahogany, it has been known as gaboon mahogany, but since it is unrelated to the true mahoganies the description is misleading and should be discontinued. The colour of the wood is fairly constant at a light pinkish-brown. It is usually straight grained with little figure, though there is a slight stripe when cut on the quarter. It weighs 430 kg/m$^3$ when dried.

## Drying
The timber dries well with comparatively little tendency to degrade during the process.

## Strength
It is a weak wood with strength properties about the same as for poplar.

## Durability
It is not very resistant to decay, but its use being mainly confined to interior work, this point is not of great importance.

## Working qualities
The timber works reasonably well with most hand and machine tools, but it is inclined to be woolly, and blunts saw teeth rather quickly. Since its use is mainly for plywood and blockboard, the processes involved in veneer production can be carried out without undue difficulty, and the resultant boards take a fine sheen when scraped or sanded.

## Uses
The most important use for gaboon is in the manufacture of plywood, blockboard and laminboard, which are used for a variety of purposes from flush doors to panelling and cabinet work. In the solid it is used commonly as a substitute for mahogany in cabinet work.

# GEDU NOHOR

*Entandrophragma angolense* C. DC.,          Family : Meliaceae
and its varieties

## Other names
edinam (Ghana); tiama (Ivory Coast); kalungi (Zaire); abenbegne (Gabon); timbi (Cameroons).

## Distribution
It occurs in the semi-evergreen forests from the Ivory Coast to Angola in the west, across to Uganda in the east.

## The tree
It is a large deciduous tree often 48m tall with an average

diameter of 1.0m to 1.5m or more. It is sometimes strongly buttressed, the winged buttresses extending as much as 6m up the trunk.

## The timber
The heartwood is typically a uniform reddish-brown, but occasional logs are much lighter coloured, sometimes a pale pink not very different from the pinkish-grey sapwood which may be 100mm wide. The heartwood darkens on exposure. The grain is interlocked, but the stripe figure produced on quarter-sawn surfaces is rather irregular and broad. It has a medium texture. The surface is lustrous, more so than sapele, and when fresh, the wood has a slight scent. It is lighter in weight than sapele, weighing about 560 kg/m$^3$ when dried.

## Drying
The timber dries rapidly with a marked tendency to distort.

## Strength
It has good strength properties for its weight, but is slightly inferior in this respect to African mahogany and sapele.

## Durability
Moderately durable.

## Working qualities
Works fairly easily with hand and machine tools, but owing to interlocked grain it has a marked tendency to tear during planing or moulding unless the cutting angle is reduced to at least 15°. It glues, nails and screws well, and takes a good polish if the grain is filled. Decorative veneer can be produced, and the plain veneer is suitable for the manufacture of plywood.

## Uses
Gedu nohor has the same uses as for sapele and mahogany, ie, furniture, interior decoration, fittings in shops, offices and ships.

# GHEOMBI

*Sindoropsis le-testui*                     Family : Caesalpiniaceae

## Distribution
Endemic to Gabon.

## The tree
A small to medium-size tree reaching a height of 15m and a diameter of 0.6m to 0.9m with a straight and cylindrical bole, usually unbuttressed.

## The timber
The sapwood is white, well defined from the red-brown heartwood which darkens on exposure. The wood contains resin canals, has a generally straight grain and a coarse, but even texture. It weighs from 650–720 kg/m$^3$ when dry.

## Drying
Dries easily and well, but logs and stock should be dipped as soon as possible to prevent losses from insects and fungi.

## Strength
Similar to dahoma in most strength categories.

## Durability
Moderately durable.

## Working qualities
Works and machines satisfactorily, and planes to a smooth surface, although the resin sometimes tends to stain the wood. Takes nails, screws, glue, stains and polish satisfactorily.

## Uses
Flooring, furniture, veneer, plywood, interior trim and joinery.

# GMELINA

*Gmelina arborea* Linn.                     Family : Verbenaceae

## Other names
yamane, yemene, gambari ; gmelina is the preferred BS name.

## Distribution
Widely distributed throughout India and Burma, the tree has been planted extensively in Africa, particularly in South Africa, Sierra Leone, Ghana, and Nigeria where it is considered one of the most important exotic trees.

## The tree
It is extremely fast growing, attaining a height of 30m and a diameter of 0.75m. The form varies with growth conditions, the best results being obtained in deep fertile, moist but well-drained soils in high rainfall districts.

## The timber
The sapwood is up to 60mm wide and only slightly distinguished from the heartwood which is creamy white or straw-coloured, often with a pink tinge. The grain is interlocked and slightly wavy, giving a stripe figure on quarter-sawn surfaces, and the texture is moderately coarse. Tension wood is sometimes present. The wood weighs about 480 kg/m$^3$ when dry.

## Drying
Air dries well and fairly rapidly, and is tolerant of high kiln temperatures.

## Strength
Although related to teak, it is inferior to that timber in general strength properties being about 17 per cent weaker in transverse strength, 30 per cent weaker in compression parallel to the grain, and some 16 per cent weaker in modulus of elasticity.

## Durability
Moderately durable.

## Working qualities
Saws easily, and has only a slight blunting effect on cutting edges. Planes and moulds to a smooth finish, but may need reduced cutting angles when knots are present. Although it moulds well, it is too soft for general turnery. Stains, polishes and glues well, and is a good peeler for veneer. Tends to split in nailing. The wood is reported to be very stable in service.

## Uses
Light construction, domestic flooring, boat building, furniture, veneer and plywood, interior joinery, carving, pattern making. Early results from African plantations showed the wood to contain a fairly high proportion of knots varying from pin knots to rather large unacceptable knots. Pruning methods have been improved, and with the normal grading for export, gmelina should prove an acceptable addition to the market.

# GREVILLEA

*Grevillea robusta* A. Cunn.                    Family: Proteaceae

## Other names
African silky-oak.

This is an introduced species which has been extensively planted in Tanzania and Kenya as a shade tree. It is a native of Australia and was formerly exported from there to the UK; in recent years supplies have been received from East Africa. It should not be confused with Australian silky-oak (*Cardwellia sublimis*).

## The tree
Grows to a height of about 30m but is generally considerably branched.

## The timber
Somewhat paler in colour than Australian silky-oak (*Cardwellia*), it is a light golden-brown, sometimes with a pinkish tinge, having a marked 'silver grain' on quarter-sawn surfaces due to the large rays. It is straight grained, and moderately coarse textured. Moderately hard, and moderately heavy, it weighs about 580 kg/m$^3$ when dried. Owing to the fact that the tree is grown in the open to provide shade on coffee and tea plantations, the wood is inclined to be relatively knotty.

## Drying
The timber needs care in drying to avoid warping and checking.

## Strength
No data are available.

## Durability
Probably moderately durable.

## Working qualities
The timber works fairly easily although some difficulty may be experienced due to crumbling of the ray cell walls. A cutting angle of 20° gives better results in planing and moulding. Nails and screws well, stains satisfactorily and takes a good polish.

## Uses
Grevillea is used for block and strip flooring; its resistance to abrasion is high. It is also used as a substitute for oak for furniture, cabinet work, shop fittings, and panelling, either in solid or veneer form.

# GUAREA

*Guarea thompsonii* Sprague and Hutch.    Family : Meliaceae
and *G. cedrata* Pellegr.

These two species show little differences in technical properties, and it is common practice to market both under the same trade name. Except in special instances this is relatively unimportant.

## Other names
*G. thompsonii*—black guarea, obobonekwi (nekwi means black) (Nigeria); diambi (Zaire).
*G. cedrata*—white guarea, obobonofua (nofua means white) (Nigeria); bossé, (Ghana); scented cedar (UK).

## Distribution
Ghana, southern Nigeria, and Liberia. It is not so common in the Ivory Coast but it is found there as well as in Gabon and Zaire.

## The tree
The trees vary in height from 15m in Zaire, to 30m or more elsewhere, with a diameter 1.0m or slightly more above the large buttresses. *G. cedrata* is usually the larger tree.

## The timber
*G. thompsonii*—Pinkish-brown, like a pale mahogany, darkening to a better colour than *G. cedrata*. Straight grain and a silky appearance. Weight about 640 kg/m$^3$ when dried.

*G. cedrata*—Pinkish-brown, with a fine texture, and cedar-like scent. Weight about 590 kg/m$^3$ when dried.

## Drying
There is little difficulty in air drying; *G. cedrata* is generally less

51

liable to split and warp. Kiln drying requires care in order to avoid exudation of a clear resin from the wood.

## Durability
Both species are considered to be durable.

## Working qualities
*G. thompsonii*—Works fairly easily with both hand and machine tools with only a slight dulling effect on cutting edges. Less woolly than *G. cedrata*, it compares favourably with the denser grades of American mahogany. When interlocked grain is present there is a tendency to pick up in planing and moulding operations and a reduced cutting angle of 20° is recommended. It finishes cleanly and takes a high polish, and also takes nails, screws and glue without trouble.

*G. cedrata*—Works fairly easily but is inclined to be a little more woolly than *G. thompsonii* and tends to dull cutting edges more readily. Cutting angles should be reduced for planing and moulding. Care is required in polishing as resin may be exuded especially in warm atmospheres.

## Uses
Both species are used for the same purposes, ie, furniture, cabinet-making, shop fitting, boat-building, high-class joinery and veneer. Because of the tendency to exude resin, care should be taken when either species is intended for such uses as instrument cases and cigar-boxes.

## IDIGBO

*Terminaiia ivorensis* A. Chev.          Family : Combretaceae

## Other names
emeri (Ghana) ; framiré (Ivory Coast). Black afara is a name used in Nigeria for the tree. As a timber name it is confusing and should not be used.

## Distribution
Occurs in Equatorial Guinea, Sierra Leone, Liberia, Ivory Coast, Ghana, southern Nigeria, in parts of the rain forest and through-out the deciduous forest areas.

## The tree
A tall tree with a buttressed trunk attaining a height of over 30m and commonly 1.0m or more in diameter. The buttresses are broad and blunt, but the bole usually is clean and straight, 20m or more above the buttress.

## The timber
A plain, pale yellow to light brown coloured wood, sometimes relieved by a zonal figure originating in the growth rings, suggesting plain oak. There is little distinction between sapwood and heartwood, though the latter is somewhat darker in colour. The grain is straight to slightly irregular, and the texture is somewhat coarse and uneven. It is soft to medium hard, and weighs about 560 kg/m$^3$ when dried. The weight is often variable, due to a prevalence of light-weight brittle-heart, particularly in large, over-mature logs. It may vary from 480 to 625 kg/m$^3$ but for general assessment, the average dry weight is as given.

## Drying
Idigbo dries readily and well, with little distortion and splitting, and shrinkage is small.

## Strength
It has excellent strength properties, being as strong and stiff as English oak in bending, although considerably softer and less resistant to shock loads. It splits easily and has been used in West Africa for roof shingles. When converting large logs the heart should be boxed out as the brittle-heart has very much lower strength properties than the normal wood. In freshly converted stock, brittle-heart may often be recognised by a distinctive pinkish colour which may develop after exposure to light for a few days. Natural compression failures, often referred to as 'thunder shakes' usually accompany brittle-heart.

## Durability
Durable.

## Working qualities
The timber works easily with most hand and machine tools. It has little dulling effect on cutting edges and a clean finish is obtained in most operations. There is a tendency however, for the

grain to pick up when quarter-sawn material is planed, and a reduction of cutting angle to 20° or less is advisable where smooth surfaces are required. Idigbo turns well and has fairly good nail and screw holding properties and will take glue well; stains effectively and reacts well to finishing treatments.

## Uses
A useful utility timber for many purposes. It can readily be converted to rotary-cut veneer suitable for plywood, and because of its stability, ease of working, durability, and attractive appearance, it is useful for fine carpentry, joinery, and construction work. It is suitable for domestic flooring, window and door frames, etc.

The timber contains a yellow colouring matter which may leach under moist conditions and is liable to stain fabrics, and it also contains tannin in sufficient quantity for the wood to become stained if in contact with iron when wet. It is slightly acidic and may tend to promote corrosion of ferrous metals. Suitable precautions should therefore be taken in those conditions of use where the wood might become moist to protect such metal as would be in contact with the wood, or by use of non-ferrous metal.


# ILOMBA

*Pycnanthus angolensis* Warb.               Family: Myristicaceae

## Other names
akomu (Nigeria); otie (Ghana); walélé (Ivory Coast); eteng (Cameroons and Gabon).

## Distribution
Ilomba is widely distributed in the rain forests of West Africa from Guinea through the Ivory Coast, Ghana, and Nigeria, eastwards to Uganda.

## The tree
It is a tall tree, attaining a height of 30m to 36m with a diameter of 0.75m or more. Buttresses are very small or absent, and the bole is straight and cylindrical.

## The timber
The wood is greyish-white to dull pinkish-brown when dry, with little distinction between sapwood and heartwood; it has a very disagreeable odour when freshly cut, but this disappears upon drying. The wood is prone to fungal staining if not converted and dried quickly. The grain is usually straight, and the texture is rather coarse. Moderately hard to soft, the wood weighs about 510 kg/m$^3$ when dried.

## Drying
Dries with a distinct tendency to split and distort.

## Strength
It has only moderate strength, and is inclined to be brittle.

## Durability
Perishable.

## Working qualities
Easy to saw, and because of the general absence of interlocked grain, relatively easy to plane. It glues and nails satisfactorily, but is difficult to polish, and absorbs too much paint.

## Uses
A light-weight timber suitable for plywood core stock, and for general carpentry. It is used as a general utility timber in its countries of origin.

# IROKO

*Chlorophora excelsa* Benth. and Hook f.      Family: Moraceae
and *C. regia* A. Chev.

## Other names
odum (Ghana and Ivory Coast); mvule (East Africa); kambala (Zaire); bang (Cameroons); moreira (Angola); tule, intule (Mozambique).

## Distribution
*C. excelsa* has a wide distribution in tropical Africa, from Sierra Leone in the west, to Tanzania in the east.

*C. regia* is confined to West Africa, where it occurs from Senegal to Ghana. There does not appear to be any significant difference between the timber of the two species.

## The tree
*C. excelsa* attains very large sizes, reaching 45m or more in height and up to 2.7m in diameter. The stem is usually cylindrical and mostly without buttresses. It occurs in the rain, and mixed-deciduous forests.

## The timber
When freshly cut, or when unexposed to light, the heartwood is a distinct yellow colour, but on exposure to light it quickly becomes golden-brown. The sapwood is narrow, being about 50mm to 75mm wide, and clearly defined. The grain is usually interlocked and the texture is rather coarse but even, and the wood weighs on average 660 kg/m$^3$ when dried. Large, hard deposits of calcium carbonate called 'stone' deposits, are sometimes present in cavities, probably as a result of injury to the tree. They are often enclosed by the wood and not visible until the time of sawing, though the wood around them may be darker in colour, thus giving an indication of their presence.

## Drying
The timber dries well and fairly rapidly, with only a slight tendency to distortion and splitting.

## Strength
Iroko has excellent strength properties, comparing well with teak, though weaker in bending and in compression along the grain.

## Durability
Very durable.

## Working qualities
Iroko works fairly well with most tools, though with some dulling effect on their cutting edges, especially when calcareous deposits are prevalent. On quarter-sawn stock, there is a tendency for grain to pick up due to interlocked grain, and a reduction of cutting angle to 15° is usually necessary to obtain a smooth surface. An excellent finish can be obtained if the

grain is filled. It takes nails and screws well, and can be glued satisfactorily.

## Uses
The timber is of great importance in both East and West Africa. It is valuable for ship and boat-building, light flooring, interior and exterior joinery, window frames, sills, stair treads, fire-proof doors, laboratory benches, furniture, carvings, marine uses such as piling, dock and harbour work, and produces a satisfactory sliced veneer.

# IZOMBE

*Testulea gabonensis*                                 Family : Ochnaceae

## The tree
The tree occurs in West Africa and is about 36m tall, with diameters ranging from 0.9m–1.2m above the thick buttresses. It has a straight, cylindrical bole, commonly from 9m to 18m in length. The distribution is scattered in the dense, mixed, equatorial forests and transitional formations, particularly in Gabon.

## The timber
The sapwood is 25mm–50mm wide, and not well demarcated from the heartwood which may be orange-yellow, grey-yellow, or pinkish-yellow, invariably with a greyish hue. The grain is straight, and the texture is very fine and even. Some cells contain a dark gum, and the wood is sometimes figured. It weighs about 800 kg/m³ when dry.

## Drying
Dries fairly well, with little distortion. The sapwood is liable to blue-stain.

## Strength
Generally similar to afrormosia in most strength categories.

## Durability
Durable.

## Working qualities
Works easily with both hand and machine tools, and finishes well in planing and moulding. Glues well, and takes stains and polish satisfactorily, and has good nailing properties, although there is a slight tendency for splitting to occur. It has good slicing properties for veneer.

## Uses
Heavy construction, flooring, furniture and cabinets, veneer and plywood, boxes for precision instruments, joinery, including doors and windows, carving, turnery and pattern making.

# KANDA

*Beilschmiedia* spp.                          Family: Lauraceae

## Other names
bitéhi (Ivory Coast) ; nkonengu (Gabon) ; bonzale (Zaire).

## Distribution
Common in the primary forest in Gabon, but widely, although sometimes scattered in almost all the tropical forests of West Africa. Gabon, Cameroons, and Ivory Coast are present producers of kanda.

## The tree
A well-shaped tree without buttresses, with a usable bole of 18m to 25m but usually producing commercial logs 6m to 7m in length, with diameters mostly between 0.7m and 0.8m.

## The timber
The sapwood varies in colour from pinkish to yellowish, and is clearly defined from the heartwood which varies in colour, according to growth conditions, from pinkish-brown to reddish-brown, to darkish brown. The grain is usually straight, and the texture is medium to fairly fine. Growth ring markings appear on plain-sawn surfaces as rather subdued, brighter lines. The wood has a dull appearance and weighs about 730 kg/m$^3$ on average, after air drying.

## Drying
Difficult and slow to dry, with a definite tendency to check

seriously, and for casehardening to develop. Requires care and protection from hot sun during air drying, followed by careful, controlled kiln drying.

## Strength
Reported to have good strength properties.

## Durability
Durable.

## Wokring qualities
Difficult to saw, and has a severe blunting effect on cutting edges due to the fairly high silica content of the wood, while gum may also contribute to this. It planes and moulds well, and is capable of a good finish, and turns well. It can be glued, stained and polished, and has good nailing and screwing properties. Although abrasive, it is said to peel and slice satisfactorily for veneer.

## Uses
Joinery, marine uses where available sizes fit the proposed use, and possibly plywood.


# LIMBALI

*Gilbertiodendron dewevrei*                    Family: Leguminosae

## Distribution
Occurs in west and central Africa where it is found in swampy forests and in dense forests with sandy soil. Also occurs along streams.

## The tree
Grows to a height of 18–36m and a diameter of 1–2m. Boles range from 6–21m in length, and are straight, cylindrical, and often without buttresses.

## The timber
The sapwood is from 50mm to 75mm wide, paler in colour than the heartwood, which varies from light brown, yellow-brown, to dark brown, often with a red tinge. Cells containing gum are

present. The grain is straight or wavy, rarely interlocked, and the texture varies from fine to coarse. Quarter-sawn stock often has an attractive figure. The wood is hard and heavy, weighing 800–900 kg/m³ when dry.

### Drying
Dries reasonably well, but with a tendency to split.

### Strength
The general strength properties are similar to those of afrormosia.

### Durability
Durable.

### Working qualities
Works fairly well, but tends to dull cutting edges fairly rapidly. It planes easily and produces a good finish. Takes paint and varnish well, and can be polished, but ample grain filling is usually required. Holds nails and screws well, but there is a tendency for thin sizes to split.

### Uses
Heavy construction, flooring, both light and heavy duty, vehicle bodies, agricultural implements, joinery, sleepers, turnery.

## LOLIONDO

*Olea welwitschii* Gilg. and Schellenb.          Family: Oleaceae

### Other names
Elgon olive (Kenya).

### Distribution
Kenya, Uganda and Tanzania.

### The tree
A large tree, up to 30m high, with a diameter of 0.75m above a buttress. Generally the tree has a straight bole.

### The timber
The timber is pale brown, often with a pinkish tinge, and

occasionally with variegated dark brown streaks giving it a very similar appearance to that of the related East African olive (*Olea hochstetteri*). The sapwood is clearly defined and is light yellowish in colour. The grain is straight or interlocked, and the texture is moderately fine. Moderately hard and heavy it has an average weight of 800 kg/m³ when dried.

## Drying
It is a slow drying species with a tendency to check and split if the drying rate is accelerated too much.

## Strength
It is reported to have good strength properties, well above average for timbers of the same weight.

## Durability
Moderately durable.

## Working qualities
The timber is fairly easy to work, both with hand and machine tools, though some dulling of their cutting edges is liable, and the grain tends to chip out in planing and recessing due to the difference in density between the light and dark zones of wood. It is liable to split in nailing and therefore requires to be pre-bored. It stains and polishes well.

## Uses
Has been used for heavy construction work, railway sleepers, and furniture in East Africa, but is generally considered a flooring timber in the UK, and for this purpose is classed as having a high resistance to abrasion.

# LONGUI ROUGE

*Gambeya africana*                    Family: Sapotaceae

## Distribution
*Gambeya lacourtiana*, or abam, occurs in the dense mixed forest all over Congo-Brazzaville and Congo-Kinshasa.
*G. madagascariensis* or famelona a grande feuilles, is found in Malagasy.
*G subnuda*, or longui noir occurs in evergreen forests of central Africa, while

*G. africana*, or longui rouge, which is perhaps the best of the species occurs also in the Congo. It should not be confused with the longui rouge of tropical America (*Chrysophyllum*).

## The tree
*G. africana* grows to a height of 21m to 30m with a diameter of 0.6m or a little more. The bole may be angular, or slightly fluted, but good specimens are straight and cylindrical, and up to 12m in length.

## General characteristics
There is little difference in colour between sapwood and heartwood, the wood being whitish when first cut, turning pinkish-buff, then olive-yellow, and finally, brownish-yellow, often with irregular dark stripes. The wood contains a pale brown-coloured gum, the grain is usually straight, but occasionally is interlocked, and the texture is fine to medium. It weighs about 710 kg/m³ when dry.

No information on drying is available, but the wood is reported to saw easily and well, to plane and mould very easily, and to produce a first-class finish. It glues well, and takes a high polish. It does not split in nailing, and holds nails and screws well, and slices and peels satisfactorily. It is moderately durable.

## Uses
Construction, domestic flooring, vehicle bodies, furniture, handles, sporting goods (the wood is said to be tough and resilient), agricultural implements, veneer and plywood, joinery, carvings and turnery.

## MAFU

*Fagaropsis angolensis* Dale.                    Family : Rutaceae

## Other names
mfu (Tanzania) ; murumu (Kenya) ; mukarakati (Kenya).

## Distribution
Found generally in the semi-evergreen forests of Tanzania and Kenya.

## The tree
A medium sized tree up to 24m high and a diameter of 0.6m.

## The timber
The heartwood is light greenish to yellow-greenish, darkening to dark greenish-grey on exposure; the sapwood is whitish. Terminal parenchyma produces whitish coloured growth lines on longitudinal surfaces. Occasionally, some logs have irregular markings which give rise to dark curls and streaks. The grain is generally straight, and the texture is medium.
Mafu is moderately hard and heavy, weighing about 672 kg/m³ on average when dried.

## Drying
Difficult to dry with a tendency to split, warp and twist.

## Strength
No information available.

## Durability
Moderately durable.

## Working qualities
The timber is easy to work and takes an excellent finish. It turns and moulds well without undue tearing of fibres, but care must be taken when working near the edges of the material to prevent the fibres breaking away. Splits rather badly when nailed if not pre-bored. Bends reasonably well, and can be peeled satisfactorily.

## Uses
Flooring, where it is considered to give a moderate to high resistance to abrasion, cabinet-making, furniture, panelling. Has been used in Ethiopia for plywood manufacture.

# MAHOGANY, AFRICAN

*Khaya* spp.                                    Family: Meliaceae

The name African mahogany covers all species of *Khaya*, although their timbers vary somewhat in character, particularly in weight. The bulk of the timber shipped is produced by *K. ivorensis* and *K. anthotheca*, each with moderately light-weight, pale to medium-red wood, and it is timber of this type

63

which is accepted commercially as African mahogany; East African *K. nyasica* is generally similar. A small proportion of *K. grandifoliola* is moderately light in weight but much of its timber and that of *K. senegalensis* is darker and appreciably heavier than that normally accepted as African mahogany. It has been suggested that such heavy wood should be marketed separately, and the name heavy African mahogany is recommended.

## Other names
*Khaya ivorensis* A. Chev. (West Africa).
*Khaya anthotheca* (Welw) A.DC. (West and East Africa).
Ghana, Ivory Coast, Takoradi, Grand Bassam mahogany, according to origin (UK); acajou d'Afrique (France); khaya (USA).

*K. ivorensis* is also known as Benin, Lagos, Nigerian, and Degema mahogany, Lagoswood and ogwango (Nigeria), and ngollon (Cameroons).

*K. anthotheca* is also known as krala (Ivory Coast); mangona (Cameroons); munyama (Uganda).

*Khaya nyasica* Stapf. ex Baker f. (East Africa).
Mozambique mahogany, mbaua, umbaua (Mozambique); mbawa (Malawi); mkangazi (Tanzania).

*Khaya grandifoliola* C.DC. (West Africa). Beninwood, Benin mahogany (Nigeria); grandifoliola (UK).

*Khaya senegalensis* (Desr.) A. Juss. (West and Central Africa) dry-zone mahogany (General); bissilom (Port Guinea); Guinea mahogany (UK).

Weight when dried (kg/m$^3$)
| | |
|---|---|
| *K. ivorensis* | 530 |
| *K. anthotheca* | 540 |

| | |
|---|---|
| *K. nyasica* | 590 |
| *K. grandifoliola* | 720* |
| *K. senegalensis* | 800 |

*Occasional pieces of light-weight *K. grandifoliola* are included with species designated as African mahogany.

## Distribution

*K. ivorensis* occurs in the coastal rain forests of West Africa from the Ivory Coast to the Cameroons and Gabon, including those of Ghana and Nigeria. *K. anthotheca* grows in West Africa in areas with lower rainfall than *K. ivorensis* requires, and is not found in the coastal belt; in East Africa it is confined mainly to Uganda and Tanzania.

*K. grandifoliola* grows at some distance from the West African coastal belt, in districts of relatively low rainfall.

*K. nyasica* occurs in East and Central Africa, particularly in Uganda and Tanzania.

*K. senegalensis* is found in the west from Senegal to Zaire and across the continent to Sudan and Uganda.

## The trees

*K. ivorensis.* Grows to a height of 30m or more with a clear bole 12m to 25m in length above the buttresses, and with a diameter of 1.0m upwards to 2.0m or more. The habit of all *Khaya* species varies considerably with the growth conditions, but the banks of rivers and streams appear to suit the requirements of the species better than drier soils. Thus *K. anthotheca* is usually not such a good shape as *K. ivorensis*, and *K. grandifoliola* is not so tall, and generally has a more crooked growth habit, though it usually attains a larger girth than other species.

*K. senegalensis* is a smaller tree and not so well shaped as the usual types of commercial mahogany. It grows mainly in the deciduous savannah forests and generally reaches a height of 15m to 24m with a diameter of about 1.0m.

## The timber

African mahogany, ie, *K. ivorensis*, *K. anthotheca*, and *K. grandifoliola* (in part).

The heartwood is distinctly pink when freshly sawn, but when seasoned varies in colour from light pinkish-brown to a deep reddish shade; the yellowish-brown sapwood is not always clearly demarcated. The heartwood of *K. grandifoliola* tends to be darker.

The grain is usually interlocked and the texture is of a coarser nature than that of American mahogany. The quality varies with the locality of growth; some localities are said to produce coarse-textured logs with spongy hearts while others are noted for the fine texture and character of their timber. A common feature is the defect known as 'thunder shake' (cross fractures), which are particularly abundant in trees with a soft or 'punky' heart.

*K. nyasica* from East Africa inclines to a reddish or golden-brown shade.

Heavy mahogany (dry-zone mahogany), ie *K. senegalensis* and *K. grandifoliola* (in part).

The timber of both these species is appreciably denser, and typically darker than ordinary commercial African mahogany, *K. senegalensis* in particular being deep red-brown with a purple tinge. In respect of grain and texture, there is little difference from the characteristics of African mahogany, but *K. grandifoliola* is reputed to be of high quality.

### Drying
African mahogany dries fairly rapidly with generally little degrade. Care should be taken to prevent distortion and splitting, and this aspect is of greater importance when drying heavy mahogany.

### Durability
All *Khaya* species are considered moderately durable.

### Strength
The strength of African mahogany compares favourably with that of American mahogany (*Swietenia*), but is more resistant to splitting. No data are available regarding strength of heavy mahogany although it can be assumed the heavier species are stronger than African mahogany.

## Working qualities

The lighter material is easy to work but the heavier species are slightly more difficult. They all have a tendency to pick up on quarter-sawn surfaces, due to interlocking grain, and a reduction of cutting angles to 15° helps to overcome this tendency. All species can be glued satisfactorily, and generally have good nailing and screwing properties. Takes a high polish and a good finish.

## Uses

African mahogany is an important timber for furniture, indoor decoration, both in the solid and as veneer, high quality joinery for staircases, panelling, and domestic flooring, boat planking and cabins, banisters and handrails.

Heavy mahogany has similar uses; *K. senegalensis* is said to provide the best surface-finishing of all the African mahoganies and is a popular timber in East Africa for lorry bodies, construction work, and decking in boats apart from the normal uses of furniture etc.

## MAKARATI

*Burkea africana* Hook.                    Family: Leguminosae

## Other names

siri (Ivory Coast); pinimo (Ghana); kola (Nigeria); mukalati (Malawi); musheshe (Zambia).

## Distribution

Widely distributed in tropical Africa, extending to South Africa, chiefly in the savannah forests.

## The tree

A fairly small tree 15m to 22m high.

## The timber

The heartwood, when first sawn, is brown with tinges of green and grey, but the colour tones down to dark brown or reddish-brown in a short time. The sapwood is paler. Very hard and heavy, but variable in weight from 750 to 1041 kg/m³ (average about 960 kg/m³) when dried. The grain is interlocked, but the texture is fine.

## Drying
Reported to dry fairly well, but with some tendency to distort and shake. A mild kiln schedule is advisable.

## Strength
A tough, strong wood, harder and stronger in bending than English oak.

## Durability
Durable.

## Working qualities
Not difficult to saw, but quarter sawn stock tends to pluck out during planing. Requires care in finishing but is capable of taking a fine polish. Should be pre-bored prior to nailing to avoid splitting.

## Uses
Heavy construction, bridges, sleepers, fencing, waggon construction, and for tool handles and flooring.

# MAKORÉ

*Tieghemella heckelii* Hutch. and Dalz.          Family: Sapotaceae

## Other names
agamokwe (Nigeria) ; baku, abaku (Ghana).

## Distribution
It occurs in Sierra Leone, Nigeria, Ivory Coast, Ghana, and Liberia, generally scattered in the moist high forest zone.

## The tree
A large tree with a straight cylindrical bole without buttresses. It attains a height of 36m to 45m and a diameter of 2.7m but

since very large trees are reported to be likely to shatter when felled, the exploitable diameter is more usually 1.0m or slightly more.

## The timber
The wood is somewhat similar to a close-grained mahogany. It varies in colour from pinkish to blood red or red-brown; the lighter coloured sapwood is usually 50mm to 75mm wide. Some logs are straight grained, but others have a striking, chequered figure and occasionally show streaks of a darker colour. The texture is much finer than mahogany, and the wood is denser, harder and heavier, weighing about 640 kg/m$^3$ when dried. The surface is distinctly lustrous.

## Drying
Makoré dries at a moderate rate and degrade is generally slight. Some distortion due to twisting may occur in some pieces, and some slight splitting tends to develop around knots.

## Strength
A tough and stiff timber comparable with American mahogany, but harder and with much greater resistance to splitting.

## Durability
Very durable. The timber is recorded as very resistant to attack by termites in Nigeria.

## Working qualities
There is a tendency for saws in particular, and other tools in general to become rapidly blunted. The blunting effect increases as the moisture content of the wood decreases, and for material with a moisture content below 20 per cent, saw teeth should be tipped with tungsten carbide. In planing, cutter angles should be reduced to 20° to avoid tearing of quarter-sawn stock. It stains and polishes well, glues excellently, but tends to split in nailing.

## Uses
Furniture, and when figured, is suitable for high-class sliced veneer. Used for doors, table-legs, chairs, superior joinery and fittings, laboratory benches, sills, thresholds and flooring, vehicle bodies, textile rollers and general turnery, cladding and panelling.

# MALANCATHA

*Malacantha* spp.           Family: Sapotaceae

## Other names
muna. mutunguru, luniondet, chepkebet (Kenya).

## Distribution
Found mainly in the mountain rain forests of Kenya.

## The tree
A large tree attaining a height of 45m with a clear bole of 27m above a heavily buttressed base.

## The timber
The timber, which has an unpleasant odour when freshly cut, is pinkish-brown in colour, often with a wavy grain which produces a fiddleback figure. The grain generally is irregular, and the texture is moderately fine. Fairly hard for its weight, which is about 496 kg/m$^3$ when dried.

## Drying
No information available.

## Strength
No information available.

## Durability
No information available.

## Working qualities
It is stated to be difficult to saw owing to the presence of silica. It planes well and takes a good finish and a high polish.

## Uses
Malacantha is a popular timber in East Africa for joinery, lorry bodies, furniture, planking and decking in boat-building, and is also used for brush backs.

# MANSONIA

*Mansonia altissima* A. Chev.          Family: Triplochitonaceae

**Other names**
ofun (Nigeria); bété (Ivory Coast); aprono (Ghana).

**Distribution**
Occurs in southern Nigeria, Ivory Coast, and Ghana.

**The tree**
A medium-sized slender tree, it reaches a height of 30m with an average diameter at maturity of 0.75m.

**The timber**
Sapwood is whitish, and the heartwood is yellowish-brown to greyish or grey-brown with frequently, a purplish tinge. Both as regards colour and grain, darker-coloured mansonia is similar to American black walnut. The colour, however, varies considerably and no strict comparison can be made between the two timbers. The grain is usually straight and the texture fine and smooth. It is fairly hard, and weighs 610 kg/m³ when dried.

**Drying**
Air dries well with little degrade but for splitting of knots and a slight tendency to warp. Kiln dries fairly rapidly and well. Shakes are inclined to extend, and also some distortion in the length may occur. Shrinkage is small.

**Strength**
Mansonia compares well with black walnut in strength, but is harder, more resistant to shock loads and stronger in bending; in other categories it is about equal.

**Durability**
Very durable. The timber is recorded as fairly resistant to termites in Nigeria.

**Working qualities**
The timber works easily with all hand and machine tools. It is better than American black walnut in cutting, and has less dulling effect on tools. It takes nails, screws, and glue well, and stains and polishes give an excellent finish.

## Uses
Mansonia is mostly used as a decorative timber for furniture and is a substitute for walnut (*Juglans* spp.) ; it is also used for cabinet making, interior joinery, pianos, and turnery.

# MISSANDA

*Erythrophleum suaveolens* Brenan.,     Family: Leguminosae
and *E. ivorense* A. Chev.
Syn. *E. micranthum* Harms.

## Other names
*E. sauveolens* and *E. ivorense:* tali (Ivory Coast) ; potrodom (Ghana) ; erun, sasswood (Nigeria).
*E. suaveolens*: munara (Uganda).
*E. ivorense*: kassa (Zaire) ; muave (Zambia).

## Distribution
The species of *Erythrophleum* occupy an extensive area in Africa. *E. sauveolens* is a mountain species of the semi-humid areas of West Africa, from which it extends at low altitudes to the boundaries of the equatorial forest. *E. ivorense* grows in the dense equatorial forest. In East Africa, both species occur in the savannah and riparian forests. They are common to tropical Africa generally.

## The tree
*Erythrophleum* grows to a height of 28m to 40m and 1.0m to 2.0m in diameter. Rounded buttresses sometimes rise rather high ; the bole is rarely very straight, and under the best conditions are seldom capable of yielding more than four commercial logs.

## The timber
The sapwood is narrow, yellowish or greyish in colour, and the heartwood varies in colour according to the locality of growth. The yellow or orange-brown with russet shading in the heartwood is a warm colour, but it darkens in time more or less depending on area of origin. The texture is coarse, and the grain is decidedly interlocked. The wood has a moderately high lustre, is moderately hard to hard, and is very heavy, weighing about 910 kg/m$^3$ when dried.

## Drying
The timber air dries slowly. There is some tendency to distort, but the wood can be dried in good condition if care is taken. Kiln drying must be conducted very slowly.

## Strength
An extremely strong timber, except in compression when it is only moderately strong.

## Durability
It is reported to be very resistant to decay, and resistant to attack by termites and teredo.

## Working qualities
A difficult timber to work and saw; planing is often difficult because of interlocked grain, but the wood turns well. The timber can take a fairly good finish and waxes and polishes well.

## Uses
Missanda is mainly used in the UK for flooring purposes; it has a high resistance to wear, and is used in schools and other public buildings. It is also suitable for heavy duty flooring in warehouses. It is an established commercial timber in the countries of origin being employed for heavy construction, exterior carpentry and joinery, for gates, decking of bridges, railway ties, harbour work.

# MOABI

*Baillonella toxisperma* Pierre.          Family: Sapotaceae
Syn. *Mimusops djave* Engl.
and *Mimusops toxisperma* (Pierre) A.Chev.

## Other names
djave (Nigeria)

## Distribution
Southern Nigeria and Gabon mostly.

## The tree
A very large tree often attaining a height of 60m and a diameter of 3m.

## The timber

The heartwood is rich red or light reddish-brown in colour; the sapwood is pale, and the texture fine. Fairly hard and heavy, it weighs about 800 kg/m³ when dried.

## Drying

Dries at a moderate rate and shows little degrade. There is slight distortion, but some twisting may occur. Treat as for makoré.

## Strength

No information available.

## Durability

Durable.

## Working qualities

Works easily but with some dulling effect on cutting edges. A cutting angle of 20° is necessary in planing quarter-sawn material. Glues and nails satisfactorily, and takes a good finish.

## Uses

Furniture, cabinet-making, flooring, decorative veneer.

# MTAMBARA

*Cephalosphaera usambarensis* Warb.    Family: Myristicaceae

## Distribution

East Africa, principally in Tanzania.

## The tree

A large tree, commonly 50m in height, with a straight, cylindrical bole, some 15m to 24m long and 1.2m in diameter above the well-developed buttresses.

## The timber

The sapwood is not clearly demarcated from the heartwood which is pale pinkish-brown with a faint orange tint, and darkening with age to a reddish-brown. The grain is usually straight, and the texture is moderately fine and even. The wood has a plain appearance, without figure or lustre, and weighs about 590 kg/m³ when dry. It is very similar in appearance to virola (*Virola* spp.) of central and south America, but is usually rather heavier than virola.

## Drying
Dries in the open air slowly but well, and also kiln-dries rapidly, but there is a tendency for thin sizes to warp, and for severe case hardening to occur, particularly in thicknesses greater than 25mm. Case hardening can be relieved however, without difficulty.

## Strength
For its weight, mtambara is about average in bending, compression, and cleavage, above average in stiffness and shear strength, but rather low in hardness and resistance to impact.

## Durability
Perishable. The wood is liable to mildew and sap stain and should therefore be dipped after conversion from the log.

## Working qualities
Very easy to work with both hand and machine tools. It can be sanded to a smooth finish, and takes stains and polish quite well. It peels easily for veneer, and takes and holds nails well. It also glues well, but phenolic resins are said to give the best results.

## Uses
Interior joinery, furniture, plywood, light construction.

# MUERI

*Pygeum africanum* Hook f.  Family: Rosaceae

## Other names
mkondekonde (Kenya).

## Distribution
Occurs in semi-tropical rain forests in Kenya, Ethiopia, Tanzania and Uganda.

## The tree
A medium-sized tree, occasionally up to 24m high, sometimes with a clear bole extending 15m above a small buttressed base. It has a diameter of about 0.5m.

## The timber
The sapwood and heartwood are not clearly differentiated when freshly cut, and are light pink in colour. Upon exposure, the heartwood darkens to light pinkish-brown. The grain is fairly straight, and the texture is medium fine, and even.

The wood is hard and heavy, weighing about 720 to 768 kg/m$^3$ when dried.

## Drying
Reported in Kenya to be a refractory species, liable to split and warp.

## Strength
A moderately strong timber, superior to English oak in all strength properties.

## Durability
Probably durable.

## Working qualities
The timber saws and works well, and planes to a smooth surface. It can be moulded and turns satisfactorily. It stains evenly and takes a high polish. Care is required when nailing to avoid splitting.

## Uses
Used in East Africa for lorry bodies, chopping blocks, bridge decking, cabinet-making and furniture.

# MUGONHA

*Adina microcephala* (Del) Hiern.          Family: Rubiaceae

## Other names
matumi, mingerhout, mugunya, watermatoemie, mowana.

## Distribution
The tree occurs in Tanzania, Mozambique, Rhodesia, Swaziland and eastern Transvaal.

## The tree
A medium to large-sized tree, with a diameter up to 1.2m. It also occurs in Kenya where it is only a shrub.

## The timber
The sapwood is not clearly defined from the heartwood which is yellowish-brown with darker markings; it darkens on exposure. When freshly cut the wood is very oily and greasy to the touch; the grain is irregular, and the texture very fine.

A hard heavy wood, it weighs from 800 to 1025 kg/m³ when dried.

## Drying
Dries fairly well, but this process should be carried out slowly in order to minimize surface checking.

## Strength
No information is available, but bearing in mind its weight and hardness it should prove to have excellent strength properties.

## Durability
Very durable.

## Working qualities
Reported to be brittle but saws fairly well. Owing to its hardness the timber is difficult to plane although an excellent surface can be obtained. It polishes well, but is difficult to glue owng to the prevalence of oil.

## Uses
Used in countries of origin for guide blocks in mining, machine bearings (because of its oily nature), waggon building and heavy construction. Has been used for flooring where it is said to equal rock maple in its resistance to wear.

# MUGONYONE

*Apodytes dimidiata* E. Mey.                    Family: Icacinaceae

## Other names
white pear, pearwood, muchai, wanda, mungaringare, tchela-laka.

## Distribution
The tree is widely distributed, and occurs in Ethiopia, Kenya, Tanzania, Central African Republic and in mountainous regions in South Africa.

## The tree
A medium-sized tree, evergreen, commonly 15m to 21m high, with a clear bole up to 15m and a diameter of about 0.5m.

## The timber
Sapwood and heartwood are not clearly defined; the wood is generally whitish to light brown with a pink tinge when freshly cut, turning greyish-brown on exposure. Generally straight grained, with a fine, uniform texture. The wood is hard and moderately heavy, weighing about 720 kg/m$^3$ when dried.

## Drying
The timber dries reasonably well, and a mild kilning schedule is recommended.

## Strength
The timber has medium strength properties, and is reported to be low in resistance to impact.

## Durability
Moderately durable.

## Working qualities
Easy to work with all hand and machine tools. Takes a good finish and a high polish. Nails and screws well.

## Uses
Used in countries of origin for constructional work, joinery, coach-building, turnery and furniture, and owing to its ease of working has been used for carving. It is reported to have a high resistance to abrasion when used for flooring. Should not be used for exterior work without preservative treatment.

# MUHIMBI

*Cynometra alexandri* C. H. Wright.          Family: Leguminosae

## Other names
muhindi (Uganda).

## Distribution
Uganda, Tanzania and Zaire. It is fairly abundant in the drier parts of tropical rain forests, but it is also found growing in swamps.

## The tree
A large tree, attaining a height of 36m and a diameter of 0.75m above the heavy plank buttresses. Trees with a larger diameter, some may be 2m across, are usually found to be unsound and hollow.

## The timber
The sapwood, which is 50mm to 75mm wide, is pale brown in colour, clearly defined from the heartwood which is light reddish-brown when freshly converted, darkening on exposure, and has irregular darker markings. A very fine textured timber, with interlocking grain; extremely hard and heavy, weighing about 910 kg/m$^3$ when dried.

## Drying
Dries slowly, especially in thick sizes, with a tendency to end splitting and severe surface checking.

## Strength
A very strong timber, twice as strong as European redwood, in bending, compression and shear.

## Durability
Very durable.

## Working qualities
Hard to work, with a fairly severe blunting effect on cutting edges. To avoid tearing of the grain, a cutting angle of 15° is recommended. Polishes well, and takes a fine finish. It needs to be pre-bored for nailing.

## Uses
Because of its high resistance to abrasion is used extensively for heavy-duty flooring, both in strip and block form. In its countries of origin it is used for heavy construction work, eg, bridge-building, mine shaft guides and also for railway sleepers.

# MUHUHU

*Brachylaena hutchinsii* Hutch.                    Family: Compositae

## Other names
muhugwe (Tanzania).

## Distribution
Found in semi-evergreen and lowland dry forests in the coastal belt and occasionally in highland forests of Tanzania and Kenya.

## The tree
A medium sized tree, up to 24m high with a diameter of 0.6m The bole is often curved and fluted and therefore it is difficult to obtain timber in large dimensions.

## The timber
The heartwood is a fairly dark shade of yellowish-brown, often with a greenish hue. The wood has a sweet scent reminiscent of sandalwood. The wood is hard and dense, generally straight grained, with a very fine, even texture. It weighs about 930 kg/m$^3$ when dried.

## Drying
The timber needs to be dried slowly and carefully to minimize hair checks and end splitting. The wood has a low movement classification.

## Strength
A strong and stiff timber, but weak in bending.

## Durability
Very durable.

## Working qualities
Due to its hardness the timber is somewhat difficult to work, and requires pre-boring before nailing.

## Uses

Since the timber is available only in short lengths at the present time its main use is for heavy-duty flooring. It has a very high resistance to abrasion, and has proved a good alternative to maple in factory floors.

In its countries of origin muhuhu is used for heavy construction, bridge decking and girders, and railway sleepers. It is exported to India where it has been used in crematoriums as a substitute for sandalwood.

# MUKULUNGU

*Autranella congolensis* A. Chev.          Family: Sapotaceae

## Other names
kungulu, kabulungu, kondo fino (Zaire).

## Distribution
Fairly widely distributed in the great equatorial forests of Africa, and is reported to be abundant in the southern part of Zaire.

## The tree
A tall tree, 30m or more high, with a cylindrical trunk free from buttresses and up to 1.0m in diameter. The trees tend to split on felling, and sometimes are defective at the centre.

## The timber
The sapwood is small and greyish, the heartwood red to dark red often marked with streaks of dark red-brown; resembles Cuban mahogany, but the colour is more variable. Hard and heavy, it weighs about 880 kg/m³ when dried. The texture is fine and the grain is usually straight but sometimes interlocked.

## Drying
No information available.

## Strength
The wood is reported to be extraordinarily tough and resistant to compression, bending, and impact loads.

## Durability
Durable, and is said to be resistant to dilute acids.

## Working qualities
Saws and planes fairly easily, takes a smooth finish and a good polish. Nails and screws fairly easily, but with a tendency to split.

## Uses
Heavy construction, marine work, bridge decking, turnery, flooring, and has been suggested for lining of acid vats by Belgian authorities.

# MUNINGA

*Pterocarpus angolensis* D.C.        Family: Leguminosae

## Other names
mninga (Tanzania) ; ambila (Mozambique) ; mukwa (Zambia and Rhodesia) ; kiaat, kajat, kajatenhout (S Africa).

## Distribution
Occurs mainly in savannah forests throughout Tanzania, Zambia, Angola, Mozambique, Rhodesia and South Africa.

## The tree
A small tree, up to 15m or slightly more, with a diameter of 0.6m. It has a short bole, usually less than 7.5m in length.

## The timber
The timber resembles other species of the *Pterocarpus* genus (padauk from Burma, Andamans and Africa) but lacks the reddish colour of padauk, being brown with irregular reddish streaks. It is also softer and lighter in weight than the padauks, weighing about 640 kg/m$^3$ when dried. The sapwood which is rather wide, is oatmeal in colour; the grain is straight to inter-locked, and the texture is medium.

## Drying
The timber has excellent drying properties both in air and kiln drying. There is only the slightest tendency for surface checking to occur. The timber dries rather slowly.

## Strength
Owing to the varying grain, even in the same log, its strength

is generally lower on average than the values for padauk, being about 30 per cent inferior in stiffness, but it is about 20 per cent more resistant to shock loads, and about equal in bending strength.

## Durability
Very durable.

## Working qualities
Easy to saw and work, although there is a tendency for inter-locked grain to pick up in planing quarter sawn surfaces; a cutting angle of 20° is therefore necessary. The wood turns well, has good nailing and screwing properties, and takes a good polish.

## Uses
Muninga is an attractive timber, suitable for panelling, high-class joinery and furniture. It makes a first-class floor with a moderate resistance to wear, and is a good timber for decorative veneer.

# MUSIZI

*Maesopsis eminii* Engl.                    Family: Rhamnaceae

## Other names
awuru (Liberia) ; esenge (Cameroons) ; muhunya (Kenya).

## Distribution
Extends from Liberia to the Cameroons, through Zaire to north-west Tanzania and into Uganda and Kenya.

## The tree
A medium sized tree in West Africa, it attains a height of 18m with a diameter of 0.5m, with a clear bole free from buttresses. It is usually much larger in East Africa, growing to 30m to 42m and a diameter of 1.0m above short buttresses.

## The timber
Heartwood olive-brown, becoming russet upon exposure, sapwood whitish to buff coloured. The grain is interlocked and

83

the texture is moderately coarse. It is light and soft but firm, weighing about 480 kg/m³ when dried. Rather attractive in appearance, with a satin-like lustre.

## Drying
Dries rapidly, but with a tendency to split and warp, and to collapse when drying thick material.

## Strength
Very similar to European redwood (*Pinus sylvestris*) in all strength properties

## Durability
Non-durable.

## Working qualities
Works easily with all hand and machine tools and finishes to a smooth lustrous surface, although there is a tendency to pick up in planing quarter-sawn surfaces. This can be overcome by reducing the cutting angle to 20°. Takes nails and screws well, but requires filling before polishing.

## Uses
Internal joinery and light construction. Should be treated with preservative for exterior purposes.

# NIANGON

*Tarrietia utilis* Sprague.                    Family: Sterculiaceae

## Other names
ogoué (Ivory Coast and Gabon); wishmore (Liberia); nyankom (Ghana).

## Distribution
Occurs in the rain forests of Sierra Leone, Liberia, the Ivory Coast and south-west region of Ghana. It is not present in Nigeria, but reappears in the Cameroons and Gabon.

## The tree
The average height is 30m with a diameter up to 1.0m; the

bole length, above the arched, plank buttresses is usually no more than 20m. The bole is cylindrical and well formed when the trees grow on well-drained sites, but in swampy areas it is twisted and irregular.

## The timber
The heartwood and sapwood are not clearly distinct. The heartwood varies from pale pink to reddish-brown; the sapwood is lighter coloured and about 75mm wide. The grain is often wavy and interlocked, so that quarter-sawn material shows an irregular stripe figure. The texture is rather coarse, and the timber has a greasy feel, due to the presence of resin. The weight is variable, from 512 to 770 kg/m$^3$, and averaging about 625 kg/m$^3$ when dried. Niangon is similar to African mahogany but coarser in texture and denser. The two timbers may also be distinguished by examination of the radial surface (quarter-sawn); in niangon the large rays are conspicuous as dark flecks; in mahogany they are hardly visible. This characteristic gives niangon an attractive figure when quarter-sawn, which is often emphasised by the interlocking grain.

## Drying
The timber presents only minor drying problems, and it dries fairly rapidly. A small proportion of the wood may show a tendency to twist. There might be slight end splitting and surface checking, and very slight collapse may occur in a few boards.

## Strength
Similar to African mahogany, but in compression, hardness, and resistance to shear and splitting, is appreciably superior and almost equal to oak.

## Durability
Moderately durable.

## Working qualities
Fairly easy to work. It does not dull cutting edges to any appreciable extent. A considerable improvement in finish is gained by reducing cutting angles to 15°. The timber stains and polishes well but requires a rather large amount of filler; excess of gum in the wood may sometimes create difficulties in

finishing. Takes nails and screws satisfactorily, and generally glues quite well, though the French recommend a preliminary treatment with a solution of caustic soda or ammonia to overcome the resinous nature of the wood prior to gluing or varnishing.

## Uses
Niangon is a general-purpose wood for carpentry, joinery and construction.

# NIOVÉ

*Staudtia stipitata* Warb.                    Family: Myristiceae

## Other names
m'bonda (Cameroons); m'boun (Gabon); kamashi, nkafi (Zaire).

## Distribution
Fairly frequent in Gabon and occasionally found in the Cameroons and Zaire.

## The tree
A fairly large tree, 22m or more in height, and up to 1.0m in diameter.

## The timber
The heartwood is red-brown to yellowish-brown with darker markings; the sapwood is pale yellowish. It is a heavy wood with a fine texture, weighing about 880 kg/m³ when dried.

## Drying
No information available.

## Strength
No information available.

## Durability
Durable.

## Working qualities
In spite of its hardness, niové is fairly easy to work, and takes an

excellent polish. Takes and holds nails fairly well. Should be quarter-sawn.

## Uses
Cabinet-making and special joinery. It is somewhat heavy to use in the solid for general work, but could be used in veneer form. It is a useful flooring wood, and is also used in Africa for canoe paddles and gun stocks.

## 'AFRICAN OAK'

*Oldfieldia africana* Benth and Hook f.    Family: Euphorbiaceae

## Other names
angouran, esson, fu, fou, esui, etu.

## Distribution
The tree has a restricted range, occurring between Sierra Leone and the western part of the Ivory Coast.

## The tree
The tree is very large, often more than 30m tall and 1.2m or more in diameter. The bole is long and clear, with low buttresses and prop roots.

## The timber
The heartwood is brown or reddish-brown, and the sapwood which is not sharply defined, is greyish-olive, sometimes with a greenish stain. The wood has a slightly bitter taste. The grain is irregular and often interlocked, and the texture is medium fine. It is a very hard, tough and strong wood, and weighs about 993 kg/m$^3$ when dried.

## Drying
No information available, but it is said to hold its place well in service.

## Strength
A very hard and strong timber with properties in all categories superior to those of ekki.

### Durability
Very durable.

### Working qualities
Difficult to work when it is dry, but it finishes smoothly.

### Uses
Used locally for heavy, durable construction, and for keelsons for boats. It is suitable for constructions subjected to water, eg bridges, bridge and other decking, floodgates.

This timber, which has no botanical connection with true oak (*Quercus* spp), was used extensively by the English and French navies some 200 years ago.

## OBECHE

*Triplochiton scleroxylon*        Family: Triplochitonaceae
K. Schum.

### Other names
obechi, arere (Nigeria); wawa (Ghana); samba, wawa (Ivory Coast); ayous (Cameroons).

### Distribution
Obeche is found in most of the countries of West Africa, especially Nigeria, Ghana, the Ivory Coast and the Cameroons.

### The tree
A large forest tree, 45m or more high with extensive sharp buttresses, rising in some trees 6m or more up the trunk. Diameter above the buttresses may be nearly 1.5m.

### The timber
The timber is creamy-white to pale yellow in colour with little or no distinction between the sapwood and heartwood; the former however, may be up to 150mm wide. It is fairly soft, but firm and fine and even in texture; the grain is often interlocked, giving a faintly striped appearance on quarter-sawn surfaces, otherwise there is seldom any decorative figure. The timber is light in weight, averaging about 390 kg/m$^3$ when dried. Brittle-heart is common in large logs.

## Drying
The timber dries rapidly and well and with little tendency to warp or shake. The timber is liable to attack by staining fungi and should be piled in stick immediately after conversion.

## Strength
Bearing in mind the light weight of the timber, obeche has good strength properties and, when compared with European redwood, is only about 15 per cent less in maximum bending strength (modulus of rupture); in stiffness it is not so good, being about 50 per cent less than redwood.

## Durability
Non-durable. The sapwood is susceptible to attack by powder-post beetles (Lyctidae and Bostrychidae).

## Working qualities
The comparative softness of the timber makes it very easy to work with both hand and machine tools. It is desirable to use sharp, thin-edged tools to avoid picking up and crumbling in cutting. An excellent finish can easily be obtained and the timber stains and polishes well but requires light filling to produce a high-grade finish. It takes nails and screws well, and has good gluing properties. Obeche peels and slices with reasonable ease.

## Uses
Interior joinery, core-stock for plywood, linings of drawers and cupboards, furniture.

# ODOKO

*Scottellia coriacea* A. Chev.          Family: Flacourtiaceae
ex Hutch. & Dalz.

## Other names
None.

## Distribution
Occurs in West Africa from Liberia to southern Nigeria.

## The tree
A slender, straight-boled tree up to 30m in height and a diameter usually not more than about 0.5m but occasionally 1.0m.

## The timber
Sapwood and heartwood are not differentiated and are whitish to pale yellow or biscuit-coloured with sporadic darker zones. It is in the beech and sycamore class, and shares many of the good properties of these timbers.

Quarter-sawn surfaces show a distinct 'silver-grain'. The grain is usually straight but is occasionally interlocked; texture is fine and uniform. It is fairly hard (slightly softer than beech) and weighs about 640 kg/m$^3$ when dried (compare beech at about 720 kg/m$^3$ and sycamore at about 630 kg/m$^3$).

## Drying
There is a tendency in air drying for odoko to check and split; stain may also develop, checks and hair-shakes may develop during kiln drying, and existing shakes tend to enlarge. Warping is not generally serious.

## Strength
Odoko has very similar strength properties to home grown beech, except that it is less resistant to shock loads.

## Durability
Non-durable, but easily impregnated with preservatives.

## Working qualities
Fairly easy to work though with some tendency to flake on quarter-sawn surfaces (owing to the silver-grain); slight brittleness may cause chipping in some operations. A very good finish is obtainable and the timber takes a good polish. It is prone to split on nailing.

## Uses
Odoko is a general utility wood for such purposes as domestic woodware, turnery, brush backs, shoe heels and as a general substitute for beech or sycamore. As a flooring timber it has a

high resistance to wear. It is used in Nigeria for cutting boards, models for casts, wooden spoons, bowls, rollers, flooring blocks.

# OGEA

*Daniellia ogea* Rolfe ex Holl.           Family: Leguminosae
and *Daniellia thurifera* Bennett.

As there are other species of *Daniellia* which should not be confused with *D. ogea* and *D. thurifera*, the distinctive name ogea is to be preferred.

## Other names
oziya, daniellia (Nigeria) ; faro (Ivory Coast) ; faro, gum copal, copal (Liberia) ; hyedua (Ghana).

## Distribution
*D. ogea* and *D. thurifera* are the most important of the numerous species of *Daniellia* that occur in West Africa. *D. ogea* occurs throughout West Africa in the drier zone of the high forest belt, while *D. thurifera* is more abundant in the wet, high forest area.

## The tree
The trees may be up to 30m high and from 1.2m to 1.5m in diameter. The boles are straight, without buttresses, and free of branches for 15m to 20m from the ground.

## The timber
The sapwood is usually wide (often from 100mm to 175mm), and greyish or straw-coloured, not sharply distinct from the heartwood which is light golden-brown to reddish-brown, sometimes marked with darker streaks. Texture is moderately coarse, and the grain is straight to interlocked ; planed surfaces have a high lustre. The timber is light and moderately soft, and weighs from 420 to 580 kg/m$^3$ when dried, the sapwood being appreciably lighter in weight than the darker heartwood.

Compression failures in the form of cross-shakes, thunder-shakes and ruptures may be prominent near the heart-centre. Vertical gum ducts are scattered among the vessels. It exudes some gum which is used for the manufacture of West African gum copal.

## Drying
Ogea dries fairly rapidly from the green with little degrade. In thick material there might be slight distortion and collapse, but the degrade is not severe. The wide sapwood quickly develops stain if not dried soon after conversion.

## Strength
The strength properties are about the same as those for abura, except in shock resistance and compressive strength, where it is slightly weaker.

## Durability
Perishable, and resistant to impregnation with preservatives.

## Working qualities
The timber is easy to work with both hand and machine tools. Interlocked grain causes tearing of quarter-sawn material in planing operations; a cutting angle of 20° and sharp, thin cutter knives are helpful in obtaining a smooth surface. The wood nails well and can be glued satisfactorily. It takes stain readily, but requires filling before polishing.

## Uses
Light joinery, boxes and crates as a substitute for spruce (it nails well), core veneer for plywood.

# OKAN

*Cylicodiscus gabunensis* Harms.          Family: Leguminosae

## Other names
denya (Ghana).

## Distribution
Okan is common in the rain forests from Sierra Leone to Liberia, the Cameroons and Gabon. It is particularly plentiful in Nigeria and Ghana.

## The tree
The tree may be 55m or more in height and 2.5m to 3m in diameter. The usual diameter of commercial logs is about

1.0m or slightly more. The buttresses are rarely more than 1.0m high, and the bole is straight, cylindrical, and without branches for about 24m.

## The timber
The sapwood is pinkish and very distinct from the heartwood which varies from yellow to brown with a greenish tinge; on exposure the colour becomes dark red-brown. The grain is interlocked and the texture is coarse, but the surface of the wood is lustrous. It is hard and very heavy, being about 960 kg/m$^3$ when dried.

## Drying
Okan tends to check and distort in drying; it dries slowly.

## Strength
Very high strength values, comparable with greenheart and karri, though slightly inferior.

## Durability
Very durable. It is recorded in Nigeria as resistant to termites.

## Working qualities
In spite of its hardness, okan works quite well, though with some dulling effect on tools, and with a strong tendency to pick up on quarter-sawn surfaces. Planing requires a cutting angle no greater than 10°. Stains and polishes quite well, but is too hard to be nailed without pre-boring.

## Uses
Piling, wharf decking, heavy duty flooring, heavy construction.

# OKWEN

*Brachystegia* spp.,                    Family: Leguminosae
including *B. eurycoma* Harms.,
*B. leonensis* Hutch. & Burtt Davy,
and *B. nigerica* Hoyle & A. P. D. Jones.

## Other names
meblo (Ivory Coast); naga (Cameroons); brachystegia (Nigeria).

## Distribution
West Africa generally.

## The tree
The different species of *Brachystegia* are generally considered as forest giants, sometimes reaching 40m in height. Its diameter may reach 2m or more, but probably most mature trees are not greater than about 1.2m in diameter. The trunk is cylindrical with winged buttresses at the base.

## The timber
The timber of the different species are similar in appearance ie light to dark brown, but light and dark alternating stripes may be present, and this feature appears to be more prominent in *B. leonensis*. Weight and workability and other properties also vary between the species.

*B. eurycoma*
Heartwood pale fawn to dark brown. Weight about 640 kg/m$^3$ when dried.

*B. leonensis.*
Heartwood light to dark brown, frequently with alternating light and dark stripes. Weight about 705 kg/m$^3$ when dried.

*B. nigerica*
Heartwood pale fawn to fairly dark brown, with occasional alternating light and dark stripes and roe figure on radial surfaces. Weight about 705 kg/m$^3$ when dried.

In all species the sapwood is wide, and up to 150mm usually. The grain is deeply interlocked, and the texture is medium.

## Drying
The wood dries fairly well, but slowly, the chief problem is distortion, but there is some tendency to end splitting and surface checking.

## Strength
Most strength properties of okwen are similar to oak, but it is harder, tougher, and more resistant to shear than oak.

## Durability
All species are probably moderately durable.

## Working qualities
Fairly easy to work with machine tools, but generally hard to work with hand tools. The deeply interlocked grain makes smooth finishing difficult in planing and moulding, and the wood is therefore not suitable for high-class finishing treatments.

## Uses
General construction that does not require high durability. Owing to its tendency to ring-shake in felling, it is advisable for conversion to be undertaken in country of origin.

Straight grained logs peel well and are technically suitable for veneer and plywood.

# EAST AFRICAN OLIVE

*Olea hochstetteri* Bak.                    Family: Oleaceae

## Other names
musheragi (Kenya).

## Distribution
It is found mainly in the rain forests of Kenya, but also occurs in Tanzania, and to a much lesser extent in Uganda.

## The tree
A medium to large sized tree, 27m in height, with an average diameter of 0.6m and has an irregularly shaped bole, rarely exceeding 10m clear.

## The timber
The heartwood is buff-coloured, attractively marked with irregular brownish, greyish, and blackish streaks, which give the wood a marbled appearance. The sapwood is pale brown without characteristic marking. The grain is slightly interlocked and the texture is very fine and even. The timber is hard and heavy, weighing about 900 kg/m³ when dried.

## Drying
The timber needs to be carefully air dried since it is somewhat refractory and liable to check and split. It dries slowly, and it is therefore advisable to protect the ends of logs and planks, and to use thin stickers. It can be kiln dried successfully provided no attempt is made to accelerate the drying rate, but internal checking or honeycombing may develop in thick material.

## Strength
The timber has excellent strength properties.

## Durability
Moderately durable.

## Working qualities
Rather difficult to work, but takes a smooth finish and turns excellently. Stains and polishes well, but requires pre-boring before nailing.

## Uses
As flooring has a high resistance to wear and is a reliable substitute for maple. Used for furniture, panelling, turnery, tool handles.

# OMU

*Entandrophragma candollei* Harms.        Family: Meliaceae

## Other names
heavy sapele (Nigeria); kosipo (Ivory Coast); atom-assié (Cameroons).

The name heavy sapele is confusing and should be discontinued; the description refers to green logs, which tend to sink in water.

## Distribution
The tree is found scattered in the dense virgin forests from

Equatorial Guinea to Zaire. It is rather rare in the western part of the area and more frequent in the Congo Basin.

## The tree
It can attain large sizes. The bole is either cylindrical or has a pronounced swelling that continues into long, ramified roots. The trunk is very straight and from 20m to 30m tall and a diameter from 0.75m to 2m.

## The timber
Sapwood and heartwood distinct; the former greyish-white to pale brown in colour, and from 25mm to 75mm wide. The heartwood resembles sapele, but is darker being reddish-brown, darkening on exposure, and usually with a purplish tinge. The texture is rather coarse and the grain is interlocked to straight. A ribbon figure is visible on quarter-sawn surfaces due to interlocking grain. The rays often contain small silica granules. When dried, omu is about the same weight as sapele, ie 640 kg/m$^3$.

## Drying
Dries rather slowly with a marked tendency to distort.

## Strength
Similar to sapele.

## Durability
Moderately durable.

## Working qualities
Works fairly readily, but is a little more resistant to cutting than sapele. Tends to tear in planing and moulding; cutting angles should be reduced to 20°. Stains readily, and polishes well.

## Uses
Since it is less attractive than sapele, its uses are more restricted to high-class carpentry, for example in naval construction. It produces quite good veneer, often with a moiré figure.

# OPEPE

*Nauclea diderrichii* (De Wild. and Th. Dur.) Merr.

Family: Rubiaceae

Syn. *Sarcocephalus diderrichii* De Wild. and Th. Dur.

**Other names**
kusia (Ghana) ; badi (Ivory Coast) ; bilinga (Gabon).

**Distribution**
This tree has a wide distribution in the equatorial forests of Equatorial Guinea, Liberia, Ivory Coast, Ghana, Nigeria and the Cameroons.

**The tree**
A large, well shaped tree, from 35m to 48m tall and 1.0m to 2.0m in diameter at breast height. The trunk is generally without buttresses, although old trees often have a basal swelling that extends not more than 1.0m above the ground. The bole is slender, cylindrical, and free of branches for 20m to 30m.

**The timber**
The heartwood is a distinctive uniform golden-yellow or orange-brown colour, clearly differentiated from the pinkish yellow sapwood which is usually about 50mm wide. The texture is coarse and the grain frequently interlocked, producing a striped or roll figure on quarter-sawn surfaces. Lustrous, very hard and moderately heavy, it weighs about 750 kg/m$^3$ when dried.

**Drying**
Needs careful drying, or checks and splits may develop. Hair-shakes often occur during drying, but warping generally is not serious. Especially in large sizes opepe dries very slowly, and it is advisable to use thin stickers when piling.

**Strength**
An exceptionally strong timber, superior to English oak in all strength categories except resistance to shock loads or splitting.

**Durability**
Very durable. Also has high resistance to marine borers.

## Working qualities
The timber works with moderate ease in most hand and machine operations and has a reasonably small dulling effect on tool edges. Quarter-sawn material tends to pick up in planing unless a cutting angle of about 10° is employed. An excellent finish can be obtained; when polishing, the grain needs considerable filling, but a high polish is obtainable. The timber tends to split on nailing.

## Uses
Piling and decking in wharves and docks, general construction, domestic flooring, waggon bottoms, sills, furniture, cabinet work, interior decoration, decorative turnery.

# OVANGKOL

*Guibourtia ehie* (A. Chév) J. Léon.　　　Family: Leguminosae

## Other names
amazakoué (Ivory Coast) ; hyeduanini, anokye (Ghana).
Note : The vernacular name hyedua is sometimes applied to this species, but more properly it refers to *Daniellia ogea*, and the term should be restricted to that tree.

## Distribution
The species is found in the Ivory Coast, Ghana, southern Nigeria and Gabon.

## The tree
Ovangkol is a tall tree, attaining a height of 45m and a girth of 2.5m ; it is buttressed to about 4m but above, the bole is usually cylindrical. On older stems there is a tendency for narrow, slightly raised horizontal rings to be formed, a characteristic shared with *Daniellia ogea*, and the reason for the erroneous application of the name hyedua.

## The timber
The heartwood is yellowish-brown to chocolate coloured, with grey to almost black stripes, and is similar to 'Queensland walnut' in appearance.

The grain is interlocked and the texture is slightly coarse. It weighs about 850 kg/m³ when dried.

## Drying
Dries rapidly and fairly well with only a slight tendency to distort, but care is needed in kiln drying thick stock in order to avoid collapse.

## Strength
No information available.

## Durability
Perishable.

## Working qualities
Ovangkol is an attractive wood, deserving more attention. It appears to be suitable for high-class furniture and joinery, flooring, and turnery, and as veneer it should provide a useful addition to the range of walnut-like woods.

# AFRICAN PADAUK

*Pterocarpus soyauxii* Taub.                    Family: Leguminosae

## Other names
camwood, barwood.

## Distribution
West Africa, particularly Nigeria, Cameroons, and Zaire.

## The tree
A medium sized tree between 15m and 30m high and a diameter of 0.6m to 1.0m. Usually has wide buttresses, and the bole is sometimes divided.

## The timber
The sapwood is of an oatmeal colour, wide, often up to 200mm, the heartwood varies from blood red to dark brown with red streaks. A hard, heavy wood, weighing between 640 and 800 kg/m³ when dried. The grain is straight to slightly inter-locked, and the texture is moderately coarse.

## Drying
Dries well, but slowly.

**Strength**
No information.

**Durability**
Very durable.

**Working qualities**
Works excellently and takes a first-class finish.

**Uses**
Although commonly known as a dye wood, it is an attractive timber suitable for furniture and cabinet-making. It holds its place well after drying and is not liable to warp. It is used for knife handles, carving, electrical fittings, paddles, oars, and agricultural implements in Africa.

# PILLARWOOD

*Cassipourea malosana* (Baker) Alston.         Family: Rhizophoraceae
Syn. *C. elliottii* (Engl.) Alston

**Other names**
ndiri (Tanzania) ; musaisi (Kenya).

**Distribution**
Fairly widely distributed throughout East Africa, mainly in Tanzania, Malawi and Kenya.

**The tree**
A tall tree, with a cylindrical bole, up to 0.6m in diameter.

**The timber**
A pale greyish or off-white coloured wood with irregular darker markings, darkening on exposure to greyish-brown or light purplish-brown. Straight grained, with a fine even texture, somewhat like birch in appearance. Moderately hard and heavy ; weight about 770 kg/m³ when dried.

**Drying**
The timber needs care, and should not be allowed to dry too rapidly, so as to avoid splitting and warping.

## Strength
Although only about 10 per cent heavier than English oak, it is nearly twice as strong in bending, and 50 per cent stronger in compression and shear.

## Durability
The timber is reputed to be durable, but not resistant to termites.

## Working qualities
Easy to saw and machine, and planes and moulds readily to a smooth surface. Takes a high polish and stains readily.

## Uses
Used in countries of origin for telegraph cross-arms, carpentry and building. May prove suitable for turnery, flooring, skis, stretcher poles, etc, or as a substitute for birch and beech.

# POGA

*Poga oleosa* Pierre.                 **Family: Rhizophoraceae**

## Other names
inoi nut (Nigeria) ; ngale (Cameroons) ; ovoga (Gabon) ; afo (Equatorial Guinea).

## Distribution
Nigeria, Cameroons and Gabon, mainly.

## The tree
A large tree, attaining a height of 45m with a diameter of 1.0m or slightly more. The bole is usually about 15m high and above this the trunk forks considerably. The tree produces edible nuts.

## The timber
The heartwood is pinkish-red, and the sapwood is narrow and white, tinged with pinkish stripes. The wood is characterised by having numerous very wide rays, giving a 'silver grain' figure when quarter-sawn, which is similar to Australian silky oak.

It is a soft wood with a coarse texture, light in weight and about 400 kg/m$^3$ when dried.

**Drying**
No information available.

**Strength**
Relatively low strength properties, and reported to be weak in bending.

**Durability**
Non-durable.

**Working qualities**
Works easily and well, and planes to a smooth surface; takes nails well.

**Uses**
Cabinet work, shop-fitting, except where strength is a requirement.

# AFRICAN PTERYGOTA

*Pterygota bequaertii* De Wild.        Family: Sterculiaceae
and *P. macrocarpa* K. Schum.

**Other names**
koto (Ivory Coast) ; kefe (Nigeria) ; awari, ware (Ghana).

**Distribution**
Found in the rain forests of Nigeria and the Cameroons.

**The tree**
A medium sized, fairly slender tree above the rather heavy buttresses. It grows to a height of 23m to 30m with a diameter of 0.5m to 0.75m.

**The timber**
The wood resembles yellow sterculia (*Sterculia oblonga*) ; there is no distinction between sapwood and heartwood, the wood being cream in colour, sometimes with a greyish tint. The grain is interlocked, and the texture is moderately coarse. Both species of *Pterygota* are lighter in weight than yellow sterculia which weighs 800 kg/m³ when dried as opposed to

*P. bequaertii* which weighs 670 kg/m³ and *P. macrocarpa* which weighs 580 kg/m³. The wood shows a fleck figure when quarter-sawn.

## Drying
Pterygota needs to be dried quickly after conversion in order to avoid fungal staining. Although it dries quite rapidly, there is a distinct tendency for surface checking to occur.

## Strength
Similar to European ash in most strength properties, but is inferior to that wood in toughness, hardness, and especially in resistance to splitting.

## Durability
Non-durable.

## Working qualities
Easy to work provided cutting edges are kept sharp; there is a tendency for quarter-sawn surfaces to tear in planing and moulding, and a reduction of cutting angle to 20° helps to avoid this. The timber can be glued satisfactorily, nailed reasonably well, and if the grain is filled can be polished to a good finish.

## Uses
Core stock for plywood manufacture, or as a backing veneer for panels. Veneer must be handled carefully since dry veneer tends to split very easily. Interior joinery, boxes, crates.

# RAPANEA

*Rapanea rhododendroides* (Gilg.) Mez.     Famliy: Myrsinaceae

## Other names
mlimangombe (Tanzania) ; mugaita (Kenya).

## Distribution
Occurs in the rain forests of Tanzania and Kenya at elevations of 1200m to 2700m.

## The tree
A small to medium sized tree up to 20m but sometimes reaching a height of 27m with a diameter of 0.5m or slightly more. The trunk is often irregular in shape.

## The timber
The wood is pinkish when freshly cut, darkening slightly on exposure. The grain is generally straight but with a tendency to spiral, and the texture is medium. The numerous large rays produce a 'silver grain' figure on quarter-sawn surfaces similar to that of Australian silky oak. Hard and heavy, the wood weighs about 910 kg/m$^3$ when dried.

## Drying
The timber should be dried slowly to avoid checking and warping.

## Strength
No information available.

## Durability
Non-durable.

## Working qualities
Fairly hard to saw, but takes a good finish and good polish. Difficult to nail.

## Uses
Cabinet-making, furniture and panelling.

## SAPELE

*Entandrophragma cylindricum* Sprague.        Family: Meliaceae

## Other names
sapelewood (Nigeria); aboudikro (Ivory Coast); sapelli (Cameroons).

## Distribution
It is found in the rain forests of West Africa from the Ivory Coast through Ghana and Nigeria to the Cameroons, and it extends eastwards to Uganda and Tanzania.

## The tree
A very large tree with cylindrical bole and small or no buttresses. Grows to a height of 45m or more, and a diameter at breast height of 1.0m or slightly more.

## The timber
The sapwood is pale yellow or whitish, the heartwood pinkish when freshly cut, darkening to typical mahogany colour of reddish-brown. Sapele is characterised by a marked and regular stripe, particularly pronounced on quarter-sawn surfaces. Occasionally mottle figure is present. It is fairly close textured, and the grain is interlocked. It is harder and heavier than African mahogany, weighing about 640 kg/m$^3$ when dried. It has a pronounced cedar-like scent when freshly cut.

## Drying
The timber dries rapidly with a marked tendency to distort. Quarter-sawn material is less liable to degrade in drying.

## Strength
Sapele is much harder than African or American mahogany, and in resistance to indentation, bending strength, stiffness, and resistance to shock loads, is practically equal with English oak.

## Durability
Moderately durable.

## Working qualities
Works fairly well with hand and machine tools, but the interlocked grain is often troublesome in planing and moulding, and a reduction of cutting angle to 15° is needed to obtain a good finish. It takes screws and nails well, glues satisfactorily, stains readily, and takes an excellent polish.

## Uses
Constructional and decorative veneer, furniture, cabinet-making, shop-fitting, boat-building, panelling, flooring, joinery.

# AFRICAN SATINWOOD

*Fagara macrophylla* Engl.                    Family: Rutaceae

*Fagara heitzii* Aubrev. & Pellegr. produces the olon of West Africa.

## Other names
*F. macrophylla* ; olonvogo, olon dur (France and Gabon).
*F. heitzii* ; olon, olon tendre (France and Gabon).

## Distribution
*F. macrophylla* has a wide range throughout West Africa from Sierra Leone to Angola, and east into Kenya, Uganda, and Tanzania. *F. heitzii* is restricted to West Africa.

## The tree
*Fagara macrophylla* is a species assuming many different forms from one end to the other of its range in the equatorial forest. It is characteristic of the secondary brush area in the Ivory Coast. where it is rarely more than 500mm in diameter. In the secondary forest of Zaire, the tree attains a height of 30m with a diameter of 1.5m.

## The timber
The heartwood and sapwood of *Fagara* spp., are not well differentiated. The sapwood is yellowish-white, and the heartwood is almost saffron yellow, sometimes veined or mottled. The grain is interlocked, and the texture is medium to fine. The wood is sweet-scented when freshly cut. *F. macrophylla* weighs from 720 to 880 kg/m$^3$ when dried, and *F. heitzii*, which produces milder timber, weighs between 560 and 640 kg/m$^3$.

## Drying
Dries fairly well; requires care if surface checking is to be avoided.

## Strength
The timber is reported to be very strong and tough.

## Durability
Durable.

## Working qualities
Moderately difficult to work, and care is needed to prevent

picking-up on quarter-sawn surfaces due to interlocked grain. Takes a fine polish.

## Uses
Could be used for furniture and joinery; it is used in its countries of origin for panelling, furniture, and cabinet-making.

# STERCULIA, BROWN

*Sterculia rhinopetala* K. Schum.          Family: Sterculiaceae

## Other names
wawabima (Ghana); aye lotofa (Nigeria); red sterculia (UK).

## Distribution
Occurs in the rain forests of West Africa from Ghana to southern Nigeria and the Cameroons.

## The tree
Grows to a height of 30m with a diameter of about 1.0m or slightly less. The tree has narrow buttresses which extend up the trunk for about 3.0m.

## The timber
The sapwood is commonly 38mm to 62mm wide, straw-coloured, and sharply defined from the heartwood which varies in colour from yellowish to a reddish, or reddish-brown. The grain is sometimes straight, but more commonly is interlocked; the texture is rather coarse. Hard and moderately heavy, it weighs about 830 kg/m$^3$ when dried.

## Drying
Dries slowly and needs care to avoid degrade, having a tendency to cup and check.

## Strength
Similar to ash, but heavier, and with the exception of resistance to shearing and splitting, is slightly stronger.

## Durability
Moderately durable.

## Working qualities
Rather woolly to work with a tendency to spring. Interlocked grain causes little tearing in planing and moulding. Tends to split when nailed and can be stained and polished satisfactorily but requires filling.

## Uses
Interior joinery and construction. In Nigeria, selected logs are used to produce veneer for plywood manufacture, and flooring blocks are also produced.

# STERCULIA, YELLOW

*Sterculia oblonga* Mast.                    Family: Sterculiaceae

## Other names
okoko (Nigeria) ; eyong (Cameroons) ; white sterculia (UK).

## Distribution
Yellow sterculia is distributed throughout the high forest zone of Nigeria and the Cameroons where it is more frequent in the deciduous forests, but the trees attain larger sizes in the rain forest.

## The tree
The trees are 24 to 30m tall and 0.75m to 1.0m in diameter. The bole is free of branches for 15m to 20m and is straight and cylindrical. Buttresses extend sharply up to about 3.5m.

## The timber
The heartwood is yellowish-white to pale yellow-brown ; the sapwood is not distinct and may be 100mm to 200mm wide. The grain is somewhat interlocked, and the texture is moderately coarse. The wood has a harsh feel. When cut on the quarter, it shows an oak-like silver grain caused by the large rays. It has a disagreeable odour when freshly cut that disappears after drying ; moderately hard and heavy, it weighs about 800 kg/m$^3$ when dried.

## Drying
Dries slowly and needs care to avoid degrade. It has a marked

tendency to surface checking and end splitting, and cupping may be a serious defect.

## Strength
Compares reasonably with ash, but is superior in bending, stiffness and crushing strength; it is appreciably weaker however, in toughness, hardness, shear and splitting strength.

## Durability
Non-durable.

## Working qualities
Easy to saw and work, but requires sharp cutting edges, and a reduction of cutting angle to 20° to avoid tearing out in planing. Stains and polishes easily, glues reasonably well, and takes nails well, but with a tendency to splitting.

## Uses
Its poor drying properties and low natural durability renders it doubtful for furniture and good-class joinery. Suitable for light interior construction.

# TCHITOLA

*Oxystigma oxyphyllum* (Harms) Leon    Family: Leguminosae
Syn. *Pterygopodium oxyphyllum* Harms.

## Other names
Iolagbola (Nigeria); kitola (Zaire); tola, tola manfuta, tola chimfuta, chanfuta (Angola).

This wood should not be confused with agba (*Gossweilerodendron balsamiferum*), also sometimes known as tola, tola branca, white tola, etc.

## Distribution
Cameroons, Zaire, and Angola.

## The tree
A large tree attaining a height of 45m and a diameter up to 2m, with a clean bole above slight buttresses.

### The timber
The timber possesses three distinct zones: i. an outside sap-wood, which is very gummy and freely exudes a copal-like, cedar-scented gum, the wood being pale yellow in colour; ii. an inside sapwood, also very gummy, but pale reddish in colour, and iii. heartwood in which the gum is present but not in such great quantity. Dark gum-rings give an almost walnut-like appearance to the timber with well-marked and striking black and pale yellow stripes. The wood is moderately heavy, weighing about 610 kg/m$^3$ when dried. The grain is variable, from straight to interlocked, and the texture is medium.

### Drying
Dries easily, but gum exudation is a problem.

### Strength
French sources of information state that mechanically it is a soft, pliable wood which, by reason of its known powers of resistance to breaking, is classed among the large number of tropical woods having a high axial compression strength, and certainly in the highest category for static flexibility. To sudden impact however, its resistance is below average. Its resistance to splitting and tension is average.

### Durability
Durable.

### Working qualities
The presence of gum chokes up tools, and hampers good finishing.

### Uses
The heartwood might be used for decorative work, but the timber is generally too gummy for general use.

## 'RHODESIAN TEAK'

*Baikiaea plurijuga* Harms.          Family: Leguminosae

### Other names
Zambesi redwood (Zambia); umgusi, mukushi, mukusi (Rhodesia).

### Distribution
Zambia and Rhodesia, in open forests, scattered, but sometimes gregarious.

## The tree
A small to medium-sized tree, up to 15m in height and 0.75m in diameter, with a clear bole varying from 3m to 4m in length.

## The timber
The heartwood is reddish-brown sometimes marked with irregular black lines or flecks, and sharply defined from the lighter coloured narrow sapwood. The grain is straight to slightly interlocked, and the texture is fine and even, giving a smooth, hard surface. The weight is about 960 kg/m$^3$ in the dried condition.
The wood should not be confused with true teak (*Tectona grandis*).

## Drying
The timber dries slowly, and with care should not warp or split excessively.

## Strength
A heavy, hard timber about 30 per cent harder than rock maple. Other strength properties have not been determined.

## Durability
Very durable.

## Working qualities
Rather difficult to work; it has an appreciable dulling effect on cutting edges. A good finish is obtained in planing and moulding if the cutting angle is reduced to 20°. The timber turns excellently, and polishes well.

## Uses
Its handsome appearance and high resistance to wear makes it an ideal flooring timber, especially for heavy-duty purposes. It is usually available in block form. It is used locally for furniture, waggon building, and sleepers.

# TETRABERLINIA

*Tetraberlinia tubmaniana*        Family: Leguminosae
J. Léon.
Syn. *Dideletia* spp.
and *Monopetalanthus* spp.

## Other names
ekop.

## Distribution
Western province of Liberia, where it is very common in the heavy rainfall areas.

## The tree
A large, tall, straight tree, without buttresses, reaching a height of 36m or more, with a diameter generally not exceeding 1.2m at maturity.

## The timber
The sapwood is light coloured with a pinkish tinge, and distinct from the reddish-brown heartwood. The wood has a lustre, is moderately coarse textured, and has an attractive grain pattern. Moderately hard, it weighs about 625 kg/m$^3$ when dried.

## Drying
No information available.

## Strength
It has excellent strength properties, comparing favourably with iroko, but somewhat superior to that timber in modulus of rupture and modulus of elasticity.

## Durability
Data are incomplete on this timber but indicate that it will be rated as **non-durable** or **moderately durable**. The Building Research Establishment, Princes Risborough Laboratory, is undertaking trials with this species which should establish the correct UK classification.

## Working qualities
No potential difficulties in planing, shaping or turning, and the wood works well with hand tools. Sliced veneer can be produced satisfactorily. The wood takes polish quite well.

## Uses
Furniture, cabinet-making, joinery, veneer.

# UTILE

*Entandrophragma utile* Sprague.     Family Meliaceae

## Other names
sipo (Ivory Coast) ; assié (Cameroons).

## Distribution
Utile has a wide natural distribution in tropical Africa. It occurs in the Ivory Coast, in the Cameroons, and in Liberia, Gabon and Uganda. The tree grows chiefly in the moist, deciduous high forest.

## The tree
The tree may be up to 45m tall and 2m in diameter above the base. The bole is straight, cylindrical, and free of buttresses, and may be 21m to 24m long.

## The timber
The heartwood and sapwood are distinct; the heartwood is pale pink when freshly cut, darkening on exposure to reddish-brown. It closely resembles the related sapele, both in appearance and properties, but is more open in texture due to the larger pores, and generally lacks the cedar-like odour of sapele. The interlocked grain produces a broad ribbon-stripe, often wider and more irregular than that of sapele. It weighs about 660 kg/m$^3$ when dried.

## Drying
Utile dries moderately slowly with a distinct tendency for distortion in the form of twist to occur, and original shakes to extend. In general however, distortion is not severe.

## Strength
Its strength properties are similar to those of American mahogany.

## Durability
Durable.

## Working qualities
Works rather well, but with a slight blunting effect on cutting

edges. A cutting angle of 15° will reduce the tendency for the interlocked grain to tear during planing and moulding. Takes stain and glue well, and polishes well after filling.

## Uses
Utile is used for the same purpose as sapele, ie, furniture, cabinets, joinery, shop-fitting, boat-building, as veneer for plywood, and for decoration.

# 'AFRICAN WALNUT'

*Lovoa trichilioides* Harms.                                Family: Meliaceae
Syn. *L. klaineana* Pierre ex Sprague

## Other names
dibétou, noyer de Gabon, eyan, dilolo (France) ; apopo, sida (Nigeria) ; bombolu (Zaire). In the USA it is known as lovoawood, tigerwood, alonawood and Congowood. It is not a true walnut ie *Juglans* spp.

## Distribution
Nigeria, Ghana, Cameroons, Zaire, Gabon.

## The tree
It is a tall tree reaching 36m to 39m in height, 1.2m diameter, having a cylindrical bole with a small buttressed or fluted base. Frequently has a clean bole of 18m or more.

## The timber
It is of a golden brown colour, marked with black streaks (caused by secretory tissue or 'gum lines'), which have given it the name 'walnut'. When planed the surface is distinctly lustrous. The sapwood is narrow, buff or light brown in colour and normally sharply defined from the heartwood, although a narrow transitional area is sometimes seen. It belongs to the mahogany family and is very similar in many respects to African mahogany. It has usually interlocked grain, giving a marked 'stripe' when cut on the quarter. It averages about 560 kg/m³ when dried.

## Drying
Its drying properties are fairly good, although existing shakes may extend slightly and some distortion occur.

## Strength
For its weight the strength of the timber is good, and is equal to American black walnut in hardness and in compression along the grain.

## Durability
It is only moderately resistant to decay, and is subject to damage by ambrosia beetles and longhorn beetles. The sapwood may be attacked by powder-post beetles (Bostrychidae and Lyctidae).

## Working qualities
It works fairly easily with most tools, but tends to pick up on quarter sawn stock due to interlocked grain, and a cutting angle of 15° should be used. Hand turning needs care, and sharp tools to avoid tearing; in the same way drills need to be sharp or the fibres will tend to tear out at the bottom of the drill hole.

The timber is fairly easy to nail, but with some tendency to split. An excellent finish can be obtained by sanding and scraping and, when filled, a fine finish can be produced.

## Uses
Furniture, cabinet-making, billiard tables, panelling, veneer, joinery, chairs, gun butts and sometimes for flooring.

# WENGE

*Millettia laurentii* De Wild.                    Family: Leguminosae

## Other names
palissandre du Congo, dikela (Zaire).

## Distribution
Mainly found in Zaire, but an associated species, *M. stuhl-mannii* occurs in East Africa. It is known as panga panga, and its general appearance and characteristics closely resemble wenge.

## The tree
Medium-sized tree, 15m to 18m in height with a diameter up to 1.0m.

## The timber

Sapwood whitish, heartwood dark brown with fine, close blackish veining, giving the wood a handsome appearance. A very hard and heavy wood, it weighs about 880 kg/m$^3$ when dried (panga panga is slightly lighter in weight at 800 kg/m$^3$). Straight grained, it has a rather coarse texture.

## Drying

Dries slowly and requires care if surface checking is to be avoided.

## Strength

The wood is stated to have good resistance to bending and to shock.

## Durability

Durable.

## Working qualities

Reported to be easy to work, but difficult to polish.

## Uses

Like panga panga, it is probably best suited to flooring, although the appearance is rather dark. Wenge produces good, decorative veneer suitable for furniture and interior decoration.

## ZEBRANO

*Brachystegia fleuryana* Chev.                Family: Leguminosae

## Other names

zebra wood (UK) ; zingana (Gabon and Cameroons).

## Distribution

Gabon and Cameroons.

## The timber

A decorative wood, light gold in colour, with narrow streaks of dark brown to almost black. The surface is lustrous, and the texture somewhat coarse. The wood is hard and heavy.

## Working qualities
Zebrano is mostly used as a veneer, usually as decorative banding. The veneer is sliced, and quarter cut in order to avoid buckling due to the alternating hard and soft grain. The veneer must be glued with care, and should be treated with clear filler before polishing. Finishes well on belt sander.

# PART II SOFTWOODS

Africa has very few indigenous conifers and these are restricted to the Mediterranean region, the high mountains of Central and Eastern Africa, and South Africa. A few genera have been introduced into East and South Africa, but of the total forest land of Africa, no more than about one per cent consists of coniferous species.

## CYPRESS

*Cupressus* spp.                              Family: Cupressaceae
mainly *C. macrocarpa* Gord.

The principle species grown in East Africa are *C. lusitanica* Mill. (*C. lindleyi*) and *C. macrocarpa* Gord., the latter species is a native of North America, and has been extensively planted in both East and South Africa. The Mediterranean cypress, *C. sempervirens* Linn. has not so far contributed to shipments of cypress to the United Kingdom which generally has been from East Africa.

### The tree
Under favourable conditions attains a height of about 30m with a bole of 0.6m to 1.0m.

## The timber
The heartwood is yellowish-brown to pinkish-brown usually distinct from the paler sapwood, which is about 50mm to 100mm wide. The grain is usually straight and the texture fine and fairly even; the growth rings are marked by a narrow band of latewood, but are not conspicuous. When dried, the timber weighs about 470 kg/m³ and has a slight cedar-like odour. It is strong for its weight, and is classified as durable.

## Working Qualities
Works readily with machine and hand tools with little dulling effect on cutting edges, but knots, usually frequent, can be troublesome. The timber takes nails well, and gives satisfactory results with the usual finishing treatments.

## Uses
Cypress is a strong durable softwood for constructional work, especially where the timber is in contact with the ground, or for external work generally.

## 'PENCIL CEDAR, EAST AFRICAN'

*Juniperus procera* Hochst.                    Family: Cupressaceae
ex. A. Rich

## Distribution
African pencil 'cedar' occurs in East Africa, mainly in Kenya, Uganda, Tanzania, and Ethiopia, in the high elevation forests.

## The tree
Generally attains a height of 24m to 30m with a diameter of about 1.5m sometimes reaching larger sizes up to 2.4m or even more. It has a tapered trunk and heavily fluted butt.

## The timber
The timber is similar to the well known Virginian pencil 'cedar' (*Juniperus virginiana*), being moderately heavy, reddish-brown in colour, fine textured and characterized by its 'cedar' scent, and fine whittling qualities. The average weight of the dried timber is about 580 kg/m³ which is slightly heavier than for the American species.

## Drying
The timber has a marked tendency towards fine surface checking during drying, especially in thick sizes which also tend to end-split. Should be considered a slow-drying timber.

## Durability
The timber is naturally resistant to decay, and is reputed to be immune to Bostrychid attack, and the heartwood to termite attack.

## Working qualities
Works easily with all machine and hand tools and normally has very little dulling effect on tool edges, but occasionally logs may have hard patches of abrasive material. An excellent finish can be obtained but cutting edges should be sharp; requires care in screwing, and is liable to split when nailed. The timber can be glued satisfactorily, and good results are obtainable with the usual finishing treatments.

## Uses
The principal use is for pencil slats. In East Africa it is used for carpentry, joinery and furniture.

# PODO

*Podocarpus* spp.                    Family: Podocarpaceae
principally *P. gracilior* Pilg.,
*P. milanjianus* Rendle,
and *P. usambarensis* Pilg.

## Other name
yellowwood (South Africa).

## Distribution
*P. gracilior* occurs at altitudes of 1200m to 2700m in Kenya, Uganda and Ethiopia, and to a lesser extent in Tanzania. *P. milanjianus* is widely distributed in Kenya at altitudes of 2100m to 3000m and is also found in Uganda, and southward through parts of Tanzania and Zambia and Rhodesia. *P. usambarensis* is found at lower altitudes in Kenya and Tanzania.

## The tree
These species attain a height of 30m or more, with diameters averaging 0.75m although *P. gracilior* is sometimes of quite large diameter.

## The timber
These species are all similar in appearance; the wood is generally a light yellowish-brown with little distinction between sapwood and heartwood. It is straight grained and of uniform texture, is non-resinous and without odour, and there are no clearly defined growth rings. The weight is similar to European redwood being about 510 kg/m³ when dry.

## Drying
Podo dries fairly rapidly with a pronounced tendency to distort, and should therefore be weighted at the top of the pile, or restrained by mechanical means in order to reduce distortion. It is also liable to split and check, and if compression wood is present, some longitudinal shrinkage can be expected.

## Durability
Non-durable and permeable. Non-resistant to attack by the longhorn beetle (*Oemida gahani*), both in the forest and in timber after conversion.

## Working qualities
Podo works easily with all hand and machine tools provided reasonable care is taken to prevent breaking out at the exit of the tool in boring, mortising, etc, because of the brittle nature of the timber. It turns, planes, and moulds to a good finish, glues satisfactorily, and takes varnish, polish and paint quite well. Some difficulty is often encountered in staining due to non-uniform penetration. It holds screws firmly, but tends to split in nailing.

## Uses
Interior joinery and fittings. *P. gracilior* is reported to be suitable for good quality plywood.

# RADIATA PINE

*Pinus radiata* D. Don                    Family: Pinaceae
Syn. *Pinus insignis* Dougl. ex Loud

## Other names
insignis (South Africa), Monterey pine (USA).

## The tree
Although the natural distribution of this species is limited to a narrow belt on the southern Californian coast, it has been widely planted in South Africa and elsewhere in the southern hemisphere. In its natural habitat it usually grows to a height of 15m to 18m but in the southern hemisphere it tends to grow fast, reaching a height of 21m to 25m in 25 to 30 years, usually with a diameter of 0.3m to 0.6m.

## The timber
The pale coloured sapwood is commonly 75mm to 150mm wide, clearly distinct from the pinkish-brown heartwood. The growth rings, although mostly wide and distinct show rather less contrast between early-wood and late-wood than those of Scots or Corsican pine, consequently the texture is relatively uniform. The average weight of the dried timber is about 480 kg/m³.

## Drying
With care the timber dries with little degrade, however where spiral grain is present, appreciable warping may occur.

## Durability
Non-durable.

## Working qualities
The timber works reasonably well and clear material has little dulling effect on cutting edges. It planes to a smooth clean finish provided cutting edges are thin and sharp. Dull, or thickened cutters tend to tear the wide zones of soft early-wood and around knots. The timber can be glued satisfactorily.

## Uses
General construction, joinery, crates and boxes, and is suitable for pulp for kraft paper.

# THUYA

*Tetraclinis articulata* Mast.                    Family: Cupressaceae

**Distribution**
North Africa and Malta.

**The tree**
A small evergreen tree or shrub of handsome pyramidal outline like *Cupressus*, occurring in North Africa mainly in Algeria and Morocco. *Tetraclinis* should not be confused with western red cedar (*Thuja plicata*). Although botanically related to both *Cupressus* and *Thuja*, *Tetraclinis* differs in that the cones only have four scales.

**The timber**
The colour of the wood is yellowish-brown red. The grain is rather soft, and the timber possesses an aromatic scent.

**Uses**
Because of the generally twisted growth characteristics, thuya is generally presented to the market in the form of burrs which are used in the manufacture of small decorative items.

# USE GUIDE FOR AFRICAN TIMBERS

## AGRICULTURAL IMPLEMENTS

afrormosia; afzelia; celtis, African (as a substitute for ash); difou; iroko; limbali; padauk, African.

## BATTERY AND ACCUMULATOR BOXES

abura; mukulungu.

## BOAT AND SHIP CONSTRUCTION

**Decking**  agba; afrormosia; gmelina; guarea; iroko; mahogany, African; malacantha; mueri; sapele; utile.
**Framing**  guarea; mahogany, African; 'oak, African'.
**Keels and stems**  afzelia; danta; 'oak, African'.
**Paddles**  niové; padauk, African.
**Planking**  agba; afrormosia; danta; gmelina; guarea; mahogany, African; makoré; malacantha; sapele; utile.
**Superstructures**  agba; afzelia; afrormosia; iroko.
**Veneers for moulding**  agba; makoré; sapele; utile.

## BOXES AND CRATES

alstonia; ilomba; ogea (substitute for spruce); pterygota; radiata pine.

## CONSTRUCTION

**Heavy**  afzelia; albizia (heavy); dahoma; difou; ekki; esia; izombe; limbali; loliondo; makarati; missanda; mueri; muhuhu; 'oak, African'; okan; opepe.
**Light**  afara; agba; akossika; albizia (light); camphorwood; gmelina; guarea; longui rouge; mtambara; musizi; niangon; ogea; okwen; radiata pine; sterculia, brown; sterculia, yellow; cypress (exterior).

## DOORS (SOLID)

abura; afara; afrormosia; afzelia; agba; camphorwood; cypress, gedu nohor; gmelina; guarea; idigbo; iroko; izombe; mahogany, African; makoré; sapele; utile.

## FANCY GOODS

bubinga; ebony; olive; ovangkol; padauk, African; 'teak, Rhodesian'; thuya (Burr); zebrano.

124

# FLOORING

abura
afrormosia
afzelia
agba
akossika
ayan
banga wanga
camphorwood
celtis, African
dahoma
danta
difou
ekaba
ekki
gmelina
gedu nohor
gheombi
grevillea
guarea
idigbo
igaganga
iroko
izombe
limbali
loliondo
longui rouge
mafu

mahogany, African
makarati
makoré
missanda
moabi
mugonha
mugonyone
muhimbi
mukulungu
muninga
niové
odoko
okan
ollem
olive
opepe
ovangkol
ozigo
panga panga
safoukala
sapele
sterculia, brown
'teak, Rhodesian'
utile
walnut, African
wenge

## FURNITURE AND CABINET WORK

abura
afara
afrormosia
afzelia
agba
akossika
aningeria
avodiré
ayan
berlinia
camphorwood

celtis, African
cordia
danta
difou
ekaba
ekebergia
ekop
ekoune
gaboon
gedu nohor
gheombi

125

## Furniture and cabinet work (cont.)

grevillea
guarea
idigbo
iroko
izombe
longui rouge
mafu
mahogany, African
makoré
malacantha
mansonia
moabi
mtambara
mueri
mugonyone

muninga
niové
obeche
olive
opepe
ovangkol
padauk, African
poga
rapanea
sapele
satinwood
'teak, Rhodesian'
utile
walnut, African

## GUN STOCKS

mahogany, African; niové; walnut, African.

## INSULATION

ceiba; erimado.

## JOINERY

**High class**

abura
afara
afrormosia
afzelia
agba
aningeria
camphorwood
celtis, African
ekaba
ekebergia
ekop
ekoune
gedu nohor
guarea
idigbo
iroko
kanda

longui rouge
mafu
mahogany, African
makoré
malacantha
mansonia
mtambara
mugonyone
muninga
niové
omu
ovangkol
poga
sapele
utile
walnut, African
cypress

**Utility**

adjouaba
akossika
albizia
alstonia
antiaris
ayan
canarium, African
ceiba
cordia
difou
gheombi
ilomba
izombe
musizi

niangon
obeche
ogea
ollem
ozigo
limbali
pencil cedar
pillarwood
podo
pterygota
radiata pine
safoukala
sterculia, brown

## LABORATORY FITTINGS
abura; afrormosia; afzelia; iroko; makoré.

## MARINE PILING AND CONSTRUCTION

**Under water**

**(a) Teredo infested waters**

afrormosia
afzelia
albizia (heavy)
ekki
esia
idigbo
iroko

makoré
muhuhu
mukulungu
muninga
'oak, African'
okan
opepe

**(b) Non teredo waters**
**In addition to the above,**
agba; dahoma; guarea; utile.

**Above water**
**(a) Docks, wharves, bridges, etc.**

afzelia
ekki
iroko
makarati

missanda
muhimbi
'oak, African'
opepe

127

## (b) Decking

afrormosia
iroko
malacantha
mueri

muhuhu
mukulungu
okan
opepe

## MUSICAL INSTRUMENTS

blackwood
boxwood
cordia
ebony, African

mahogany, African
mansonia
sapele
utile

## PATTERNMAKING

abura; mahogany, African; izombe

## SHOP FITTINGS

abura
afara
afzelia
agba
avodiré
danta
gedu nohor
grevillea
idigbo

iroko
mansonia
niangon
opepe
padauk, African
poga
sapele
utile
walnut, African

## SILLS AND THRESHOLDS

afrormosia
afzelia
agba

dahoma
makoré
opepe

## SPORTS GOODS

agba
berlinia
celtis, African
danta
ebony, African
ekki
gaboon
guarea

longui rouge
mahogany, African
obeche
padauk, African
podo
sapele
utile

## STAIR TREADS

afrormosia
afzelia
ekki

iroko
opepe

## TOOL HANDLES

blackwood (knife)
celtis, African
danta

makarati
olive
padauk, African (knife)

## TOPS FOR COUNTERS

afrormosia
afzelia
guarea
iroko

mahogany, African
sapele
utile

## TURNERY

abura
blackwood
boxwood
bubinga
danta
difou
ebony, African
ekaba
ekoune
guarea
iroko
izombe

limbali
longui rouge
mahogany, African
makoré
mansonia
mugonyone
mukulungu
odoko
olive
opepe
ovangkol
padauk, African

## VEHICLE—BODY WORK

abura
agba
albizia
ayan
berlinia
dahoma
danta
iroko

limbali
longui rouge
makarati
makoré
malacantha
mueri
mugonyone
opepe

# VENEER AND PLYWOOD

**Corestock**

akossika
antiaris
canarium, African
ceiba
gmelina
igaganga

ilomba
kondrotti
obeche
ogea
ollem
pterygota

**Decorative**

afara
afzelia
aningeria
avodiré*
bubinga
difou
ebony, African
ekaba
ekoune
gheombi
grevillea
guarea
iroko
izombe

lingui rouge
mahogany, African
makoré
moabi
niové
omu
ovangkol
sapele
tetraberlinia
thuya burr
utile
walnut, African
wenge

**Utility†**

abura
adjouaba
afara
agba
akossika
aningeria
avodiré
difou
ekoune
gaboon
gheombi

idigbo
igaganga
mafu
mtambara
okwen
ozigo
podo
pterygota
safoukala
sterculia, brown

* ex selected figured logs

† utility veneer for plywood manufacture, with or without decorative face veneer, chip baskets, and small laminated articles.

# AMENABILITY OF HEARTWOOD TO PRESERVATIVE TREATMENT

## Extremely resistant

afrormosia
afzelia
albizia
avodiré
banga wanga
blackwood
camphorwood
canarium, African
difou
ebony, African
ekki
esia
gedu nohor
guarea
iroko

kanda
limbali
mahogany, African
makoré
mansonia
moabi
muhimbi
niangon
okan
okwen
sterculia, brown
sterculia, yellow
'teak, Rhodesian'
utile
'walnut, African'

## Resistant

adjouaba
agba
ayan
berlinia
cordia
dahoma
danta
ekaba

gheombi
igaganga
izombe
muninga
obeche
omu
ozigo
sapele

## Moderately resistant

abura
afara
celtis, African
ekoune
gmelina

longui rouge
mtambara
ogea
opepe
safoukala

## Permeable

akossika
alstonia
aningeria
antiaris
bombax

ceiba
ilomba
musizi
odoko
pterygota

# AMENABILITY OF HEARTWOOD TO PRESERVATIVE TREATMENT

The above classification refers to the ease with which a timber absorbs preservatives under both open-tank (non-pressure) and pressure treatments. Sapwood, although nearly always perishable, is usually much more permeable than heartwood, accordingly, the above classification refers to the relative resistance of heartwood to penetration.

### Extremely resistant

Timbers that absorb only a small amount of preservative even under long pressure treatments. They cannot be penetrated to an appreciable depth laterally, and only to a very small extent longitudinally.

### Resistant

Timbers difficult to impregnate under pressure and require a long period of treatment. It is often difficult to penetrate them laterally more than about 3mm to 6mm.
Incising is often used to obtain better treatment.

### Moderately resistant

Timbers that are fairly easy to treat, and it is usually possible to obtain a lateral penetration of the order of 6mm to 18mm in about 2-3 hours under pressure, or a penetration of a large proportion of the vessels.

### Permeable

Timbers that can be penetrated completely under pressure without difficulty, and can usually be heavily impregnated by the open-tank process.

# RESISTANCE TO MARINE BORERS *

## Very durable

afrormosia
afzelia
albizia
ekki
esia
iroko
makoré
missanda

muhuhu
mukulungu
muninga
oak, African
okan
opepe
padauk, African
teak, Rhodesian

## Moderately durable

agba
ayan
banga wanga
berlinia
dahoma
danta

guarea
idigbo
malacantha
mansonia
utile
walnut, African

## Non-durable

abura
afara
alstonia
antiaris
bombax
camphorwood
canarium, African
ceiba
celtis, African
cordia
gaboon
gedu nohor
ilomba

loliondo
mahogany, African
mugonha
muhimbi
musizi
obeche
odoko
ogea
okwen
omu
pillarwood
pterygota
sapele

*Marine borers; This classification is based mainly upon TRADA exposure trials at Shoreham; durability should be interpreted as follows:—

Very durable  Suitable under conditions of heavy attack by *Teredo* and *Limnoria*.

Moderately durable  Suitable under conditions of moderate attack, mainly *Limnoria*.

| Non-durable | unsuitable, or suitable only for short service life. |

For further information see TRADA publication, 'Timber for marine and fresh water construction'.

## TERMITE RESISTANCE (HEARTWOOD) *

Termites, (order Isoptera). The classification is based upon the reputed resistance to attack by both dry-wood and subterranean termites. Where the resistance to either type differs, the lower rating is given.

**Very resistant**
makoré, muhuhu.

**Resistant**
afzelia
agba
albizia
E. A. camphorwood
danta
ekki
esia
iroko
izombe
kanda
missanda
muhimbi
mukulungu
muninga
okan
okwen *(B. leonen — sis)*
olive
opepe
padauk, African
panga-panga
sterculia-brown

*Resistant (cont)*
'teak, Rhodesian'
wenge

**Moderately resistant**
ayan
berlinia
dahoma
ebony, African
ekaba
gaboon
gedu nohor
gmelina
guarea
idigbo
limbali
longui rouge
mansonia
omu
sapele
'walnut, African'

**Non-resistant**
abura
afara
akossika

*Non-resistant (cont)*
alstonia
aningeria
antiaris
bombax
canarium, African
ceiba
celtis, African
cordia
*Dacryodes* spp.
difou
ekoune
gheombi
ilomba
makarati
musizi
obeche
odoko
ogea
okwen *(B. euryco — ma)*
pillarwood
pterygota
rapanea
sterculia, yellow
utile

# 2
# SOUTH AMERICA

VENEZUELA

SURINAM

FRENCH GUIANA

GUYANA

COLOMBIA

ECUADOR

PERU

BRAZIL

BOLIVIA

PARAGUAY

Pacific Ocean

URUGUAY

CHILE

ARGENTINA

Atlantic Ocean

# INTRODUCTION

*The scope of this chapter is confined to the continent of South America which forms the southern part of the land of the Western Hemisphere. It is a distinct land-mass, surrounded by sea, except where the Isthmus of Panama (pierced by a canal) joins it to the Northern continent.. The east coast is washed by the Atlantic, the west coast by the Pacific, while the north coast borders the Caribbean Sea.. The south of the continent extends into the Antarctic Ocean.*

*Its geographical relation to North America somewhat resembles that of Africa to Europe. Brazil, occupying about half of the continent, is Portuguese in origin and language, while the other republics are Spanish in speech and origin.*

*While the geographical scope of the chapter is thus confined, the various timber species described in it are not similarly limited by geographical boundaries; in some cases a given species may be widely distributed and its habitat extend into Central America and the West Indies, while in other instances different species of an important genus may be commercially of greater interest in one trade area than in another.*

*The current shortage of quality virgin timber has induced the timber and associated industries of the world to look increasingly for additional sources to help close the supply gap, and tropical American timber species in particular have so far been rather ineffectively used, primarily because of insufficient knowledge of the physical characteristics and properties of many of the timbers.*

*It is essential, therefore, that the qualities of potentially useful timbers be thoroughly examined, and in order to present as wide a coverage of information as possible it has been thought necessary to publish this chapter on the timbers of South America in addition to the chapter on timbers of Central America and the Caribbean which appears in Volume 2.*

*In this way it is possible to separate a widely distributed species into a particular area of greater commercial emphasis. For example, some 80 species of the genus Eschweilera occur over an area extending from Tinidad to Brazil, but only about two of these species are* presently of commercial interest, and the timber they produce, manbarklak, is concentrated commercially, at least for the present, on Guyana and Surinam. Accordingly, this timber is described in this publication and not in its companion volume, although notes as to the full distribution of the genus are given.

Furthermore, while British Standard names are used as far as possible as the main heading to an entry, in some instances the preferred local name is used, generally because it is more readily understood in local negotiations,but more specifically because shifting spheres of interest are presently concentrated in one area more than another. For example, 'crabwood of British Guiana' is a description that has been known and understood for many years, and while the now Guyana can still supply this wood, the sphere of trade interest has shifted, or extended to Brazil where it is better known as andiroba.Therefore it is under that heading that the wood is described in this chapter, but the name crabwood has not been ignored.

Every endeavour has been made to avoid confusion between the many duplications of a vernacular name applied to several, different species, a peculiarity which characterizes tropical American timber. It is hoped that the information offered here and in Volume 2 will help all who deal in timber to determine the potential value of the species described for any immediate use.

# FORESTS OF SOUTH AMERICA

South America, extending through every habitable latitude from the North Tropics (12° N) to the Antarctic (56° S), possesses nearly every climate and almost every possible variety of soil, elevation, scenery and products, both vegetable and mineral. Despite this infinite diversity, the continent is shaped by nature with a certain vast simplicity, being broadly divided into a region of mountains, a region of tropical forests, and a region of open plains. Firstly, there is the huge chain of the Andes, secondly, the boundless Amazonian forests, and thirdly, the wide plains of the Argentine. The main feature of the continent is the gigantic Cordillera of the Andes, which stretches from the north of Colombia to the south of Tierra del Fuego, some 7000 kilometres.

Clinging closely to the Pacific, this gigantic water-shed throws the bulk of its waters eastwards into the colossal water system of the three great rivers, Orinoco, Amazon, and Paraná. The trade winds from the Atlantic sweep across the tropical parts of the continent, strike the east Cordillera, and shed their moisture on the soaking forests, through which wind the multitudinous upper waters of the great rivers. A striking contrast appears on the steep west descent of the Andes. In the south tropics and sub-tropical parts, the coastal strip between the west Cordillera and the Pacific forms a rainless desert 1600 kilometres long. North and south of this desert the Pacific coast presents a very different character. To the north, the coastal strip widens out into the well-watered tropical forests of Ecuador, and then narrows north along the damp, hot shore of Colombia.

To the south, there stretches the fertile, temperate region of central Chile. Farther south, the perpetual wet winds from the Pacific beat upon the moist forests of southern Chile.

East of the southern Andes stretch the wide tree-less plains of the Argentine pampas, and to the north, this open plain merges into the sub-tropical and tropical forests which fill the great river systems of the central part of the continent.

It has been estimated that the forests of South America cover approximately 44 per cent of the land area. They consist principally of tropical hardwoods (89 per cent), temperate hardwoods (6 per cent), and conifers (5 per cent). Two types of tropical hardwoods occur, the most abundant are those found in the dense, humid rain forests which characterizes Guyana, Surinam, French Guiana, and the great Orinoco and Amazon River basins, and which occur along the eastern coast of Brazil. This type of forest is noteworthy for the great number of species, and for the size and frequency of the individual trees. It has been estimated that there are over 2500 different tree species in the Amazon forest alone.

In the drier parts of Brazil and the Argentine an open, deciduous type of forest occurs, while mixed forests of conifers and temperate hardwoods occur along the northern Andes, and again in the southern Cordilleras. Conifer forests, consisting chiefly of Parana pine (*Araucaria angustifolia*) cover large areas of southern Brazil and the Argentine, while similar areas, with Chile pine (*Araucaria aracana*) the dominant species, occur in Chile. In Paraguay and the Argentine there are extensive areas of open forest with quebracho (*Schinopsis* spp.) representing the world's most important source of tannin.

**Forest of the Amazon**
The greatest water system and also the greatest virgin forest-land, lie mainly within Brazilian limits. The country is mostly contained within the tropics, yet the great plateau south of the Amazonian valley provides almost a sub-continent of temperate uplands in tropical latitudes. The internal navigable waterways are some 100 000 kilometres in length, with the most stupendous feature being the forest-clad valley of the Amazon.
This forest stretches some 3200 kilometres from east to west. In its broadest part, in the far interior, it is about 1600 kilometres wide, and nearer the Atlantic it narrows to about 400 kilometres. The tree canopy, the thick and luxurious foliage, and the intertwined lianas create an artificial darkness, excluding the sun's rays but not its heat. For centuries the forest teemed with life; animals, reptiles and birds flourished without intrusion of the European, but progressively, the obstacles to economic development, scanty population, scarcity of labour, the tendency to conglomeration in the big cities, and high cost of living, are being overcome, and the wealth of the forest, its varied and beautiful timber, its dye-woods, its fibres, its

medicinal or nutritive plants, fruits, and nuts, and its cellulose, is being more and more exploited, and man has begun to replace the long secluded dominion of beasts, reptiles, and birds.

A Brazilian statesman once remarked that Brazil's greatest assets were her greatest obstacles, namely the rivers and hills. The overcoming of these obstacles has opened up a boundless economic prospect. The unwholesomeness of the river valleys now yields to modern science, and the cost of the labour needed to penetrate the heights by road, rail, and the adaptation of water power is being repaid by the world's interest in Brazil's resources.

From a timber point of view, Brazil represents a teeming source of supply, and already more than 200 sawmills are operating in Amazonia together with veneer and plywood mills. In 1973 Brazil harvested 23 800 000 cubic metres of timber, almost double that of any other tropical country in the world.

The timbers described in this booklet include some that are common and readily available, and others that are common to local use but potentially of interest to the UK market. In some cases these species may be of limited distribution, or of limited availability due to local demand.

# PART I HARDWOODS

## ACAPU

*Vouacapoua americana* Aubl.         Family: Leguminosae

**Other names**
wacapou, épi de bleu, bois angelin (French Guiana) ; bruinhart, wacapoe (Surinam).

**Distribution**
Acapu is a very important timber tree in Brazil, attaining its best development in the State of Para, but apparently it does not extend westwards into the State of Amazonas. It is also common in French Guiana and Surinam, although in this latter area, large-size trees are becoming scarce in the presently accessible forests.

**The tree**
Tall, slender and unbuttressed, the trees attain a height which usually produces a clear bole 15.0m to 22.0m long, with a diameter of 0.6m.

**The timber**
A hard, heavy, dense wood ; the heartwood is dark olive to dark chocolate in colour, and is clearly demarcated from the cream coloured sapwood which is 18mm to 32mm wide. Numerous fine lines of parenchyma which are initially lighter brown in colour, but eventually turn to almost black, make the wood unusually attractive. The grain is straight to slightly roey, and the texture is uniformly coarse. The wood weighs about 960 kg/m$^3$ when dried.

**Note**
*V. pallidior* also occurs in Brazil, and is also called acapu. It is however, lighter in colour, being dull chestnut-brown with narrow streaks of dark brown and fine pencil stripes of lighter brown.

**Drying**
Dries at a moderate rate with slight warping in the form of cupping and twisting. Slight end splitting and surface checking, and casehardening are liable.

## Strength

Acapu has excellent strength properties. It is superior to white oak (*Quercus alba*) in bending and compression parallel to the grain, shock resistance and side hardness, but inferior to that timber in compression and tension across the grain, in cleavage resistance, and very slightly in shearing strength.

## Durability

Very durable.

## Working qualities

Moderately difficult to work. Smooth surfaces are obtained in sawing and planing, but the coarse grain causes some rough and torn grain in boring and mortising.

## Uses

Acapu is one of the most highly regarded woods of Brazil, Surinam, and French Guiana. It is used for furniture, cabinet making, flooring, general construction, and railway sleepers, in which use, untreated acapu has remained sound for 6 to 8 years in poorly drained soil in Brazil, and for 18 to 20 years in well-drained soils. Similar life of the untreated wood used in South Africa has also been experienced. Its natural resistance to marine borers make it suitable for piling and other marine work.

# ANDIROBA

*Carapa guianensis* Aubl.                    Family : Meliaceae

## Other names

crabwood (Guyana); krappa (Surinam); tangare (Ecuador). Also known variously as 'Brazilian', 'Guyana', 'Demerara', 'Para', and 'Surinam mahogany'. These names are confusing however, and should be discontinued.

## Distribution

Occurs in Guyana, Surinam, French Guiana, Colombia, the overflow delta lands of the Orinoco in Venezuela, Peru, and is very abundant in the Amazon flood plains in the States of Para and Amazonas of Brazil. It is also found in the West Indies and Central America.

## The tree
The tree is evergreen, 30.0m or more high, with a diameter of 1.0m although on good sites larger trees occur. The tree is straight and of good form, with boles 9.0m to 27.0m long, according to site conditions, above the short buttresses or swollen bases. The trees grow rapidly and reach felling size in Guyana in about 25 years in the marsh forests, around 30 to 35 years in the mora forests, and in about 40 to 60 years in the hill forests.

## The timber
Andiroba is closely related to the mahoganies (*Swietenia* and *Khaya*), and resembles mahogany and some grades of cedar (*Cedrela*) in colour, general appearance, and technical properties, but lacks the high lustre and attractive figure present in the better grades of mahogany. The heartwood is a light salmon or pale pink to reddish-brown when freshly cut, becoming reddish-brown to brown when dry. The general colour is somewhat darker than mahogany because of the accumulated dark-coloured gum in the vessels.

The sapwood is pinkish at first, turning pale brown or greyish, often with brown or black flecks when dry. It is not sharply defined from the heartwood, and is 25mm to 50mm wide.

The grain is usually straight, but interlocked grain and occasional fiddleback mottle occurs in larger logs. The texture varies from coarse to fine, but is mostly medium in texture.

There is some variability in texture, weight, hardness, and colour, according to site conditions ; wood from swamplands is softer, lighter in colour, more coarse, often woolly, and floats high in water, indicating lighter weight, while wood from hillsides is darker, has interlocked grain, is heavier and denser, but wood from the same site and even the same tree may vary in colour. Ripple marks occur sporadically in the denser tissue.

The weight of the wood varies from 576 to 736 kg/m³, but on average is considered to be about 640 kg/m³ when dried.

## Drying
Andiroba dries rather slowly with a tendency to distort and split. The wood varies considerably in its rate of shrinkage when dried from the green state, the ratio of tangential to radial shrinkage being relatively high, indicating non-uniform shrinkage in these two directions, and suggesting extreme care is

essential in the initial drying process. The movement value however of the wood in service compares favourably with Honduras mahogany, and quite well with African mahogany as the following figures show.

Andiroba                    1.5 per cent tangentially;
                            1.3 per cent radially.

Honduras mahogany 1.3 per cent tangentially;
                            1.0 per cent radially.

African mahogany        1.3 per cent tangentially;
                            0.8 per cent radially.


## Strength
The mechanical properties of andiroba vary according to density and to some extent with the source of the material. Princes Risborough Laboratory rated air dry wood about 30 per cent stronger in bending and resistance to suddenly applied loads, about 40 per cent more resistant to splitting, 50 per cent stiffer, and about 60 per cent harder on the side grain than Honduras mahogany, while Yale University rated the wood superior to mahogany (*Swietenia*) in all properties other than work to maximum load and shear.


## Durability
Moderately durable, and comparable to mahogany in weathering properties.


## Working qualities
More difficult to machine than mahogany, but can be worked quite easily with machines and hand tools. It has good planing, moulding, shaping, turning, mortising, sanding, and boring properties. Straight grained material machines smoothly, but quarter-sawn stock is apt to pluck out when wavy grain is present. A cutting angle of 15° is helpful in reducing the amount of sanding sometimes needed to obtain a smooth finish. Although tending to split when nailed, it holds screws better than Honduras mahogany. It glues well, and takes all kinds of finishing treatments very well but softer varieties require the use of filler. It peels well for veneer, but logs have a distinct tendency to split at the ends causing a certain amount of loss.

## Uses

Andiroba is used in the countries of origin for furniture, turnery, shingles, interior trim, flooring, boxes and crates, and in house construction for rafters, studding, sheathing, louvres, weather strips, sash-frames, and is one of the most preferred substitutes for mahogany.

It has been used successfully in the Netherlands in veneer form as a substitute for oak, and at one time was used in the German aircraft industry for plywood manufacture.

## ANGELIM PEDRA

*Hymenolobium* spp.                    Family: Leguminosae

### Other names

sapupira amarella, angelim do Para (Brazil).

### Distribution

The centre of distribution is the upland rain forests of the central and eastern parts of the Brazilian Amazon region, extending northwards into the Guianas and Venezuela and southward to Rio de Janeiro.

### The tree

*H. excelsum* Ducke appears to be the principal species, although some dozen species occur in the areas mentioned. They are all fairly large trees, on occasions being extremely large, sometimes 45m tall, and 3m in diameter.

### The timber

Angelim pedra has many features in common with *Andira* spp. which produces angelim or partridge wood of the Caribbean and central and tropical America.

The sapwood of angelim pedra is whitish or greyish in colour, 25mm to 50mm wide, gradually merging into the heartwood which is pale brown when fresh, darkening on exposure to accentuate the differences between the darker fibre layers and the lighter parenchyma bands and stripes. The wood has a small figure of the partridge-wing type, but since this is rather sub-dued, it cannot be considered a specially decorative feature. The wood is hard, tough, strong, and heavy, weighing from

704 kg/m³ to 993 kg/m³ when dry. In some pieces the weight exceeds the latter, due to small patches in the wood containing heavy deposits of hard gum in the cells.

## Durability
Very durable.

## Working qualities
Saws and works fairly well, but is rather difficult to plane due to the alternating bands of hard and soft tissue and sharp cutting edges are essential to obtain a smooth finish. Takes nails and screws well.

## Uses
Constructional purposes, particularly for marine applications. *H. excelsum* was classed ninth in marine borer resistance, of 37 timbers tested at Wrightsville, USA, greenheart ranking twelfth in these tests.

# ARARIBA

*Centrolobium* spp.                  Family : Leguminosae

## Other names
arariba amarelo, a. carijo, a. rajado, a. rosa, a. tinga, a. vermelho, oleo amarelo (Brazil) ; amarillo de Guayaquil, amarillo lagarto, canario (Ecuador) ; balaustre (Colombia and Venezuela).
In the markets of Rio de Janeiro and Sao Paulo, the general term for the woods of the genus *Centrolobium* is arariba, with qualifying adjectives which describe the colour of the wood. eg 'amarelo' (yellow), 'branca' (white), 'rosa' (rose or pinkish), and 'vermelho' (red). Thus, the timber known on the UK market as amarillo, generally the product of *Centrolobium ochroxylon*, is probably better known in Brazil as arariba amarelo. It should be noted that both common names, ie amarillo and amarelo, denote a yellow wood, but the botanical classification 'ochroxylon' also means 'yellow'.

## General characteristics
The sapwood is creamy-white, sharply demarcated from the heartwood, which, according to species varies from yellow

through orange to red, but all are generally variegated with red or purple-black streaks. The wood is generally of medium texture, while the grain varies from straight to irregular. The weight of the various species is from 752 to 1000 kg/m³ when dried.

The woods work well with machine tools, but the blackish zones tend to be hard and brittle and liable to break out in cutting and planing, especially with hand tools.

## Uses
Arariba is used in the countries of origin for civil and naval construction, joinery, doors, flooring, tight cooperage, and selected logs are said to produce veneer for furniture and panelling. The various species are all reputed to be very durable.

# BAGUACU

*Talauma ovata* St. Hil.                    Family: Magnoliaceae

## Other names
magnolia do brejo, uvaguacu (Brazil).

## Distribution
Brazil, where it occurs in Minas Gerais, Rio Grande do Sul, and most frequently in the Estados de Parana and Santa Caterina.

## General characteristics
The leaves, flowers and fruit of *Talauma* are typical of the North American magnolia, while its wood is not unlike the further related *Liriodendron tulipifera*, which produces the yellow poplar of North America, being almost white in colour with a ring-growth figure appearing on flat surfaces. The wood is straight grained, of medium texture, and possesses a superficial lustre. It is non-siliceous, and weighs about 600 kg/m³ when dried.

It is reputed to dry easily with little degrade; shrinkage values in drying from the green to 15 per cent moisture content are, radial, 3.9 per cent; tangential, 9.3 per cent; volumetric, 14.4 per cent.

## Uses
It is a non-durable wood, fairly easy to work, and is used in Brazil for carpentry and joinery, interior partitioning.

# BALSA

*Ochroma pyramidale* Urb.,                    Family : Bombacaceae
Syn. *O. lagopus* Sw., *O. bicolor* Rowlee

## Other names
guano (Puerto Rico and Honduras) ; lanero (Cuba) ; polak (Belize and Nicaragua) ; topa (Peru) ; tami (Bolivia).
Balsa is Spanish for 'raft', developed when the early Spanish colonists observed Indians using the wood in the construction of rafts.

## Distribution
Widely distributed in tropical America, its natural range extends from Cuba to Trinidad and on the continent from southern Mexico through Central America into Brazil, Bolivia, Peru, Ecuador, and Venezuela. In Central America, the tree grows in Belize, Guatemala, Honduras, Nicaragua, Costa Rica, and Panama. It has also been introduced into India and Indonesia.

## Note
The bulk of the world's supply of balsa is grown in Ecuador, where the rich soil and high temperature and rainfall form ideal growth conditions.

## The tree
Balsa is a rapid growing tree with a very short life. It often attains a height of 21.0m and a diameter of 0.5m in 7 years, and on good sites slightly larger dimensions in 5 years. After 8 years, the heart tends to develop a pink colour which is inferior to the sapwood. The trees reach maturity in 12 to 15 years, after which they deteriorate rapidly, growth slows, the heart-wood becomes waterlogged and doty, and the new growth is hard and heavy.

## The timber
Balsa is the lightest and softest timber used commercially. It possesses an unusually high degree of buoyancy and provides very efficient insulation against heat and sound ; where these properties are essential, the wood is adaptable to a great number of special uses.
The best balsa is almost white in colour with a lustrous surface ;

149

the grain is open and straight. It varies greatly in weight, some-times from 120 kg/m³ at the centre of the log to 340 kg/m³ near the outer edge. However, wood cut for export generally weighs from 128 kg/m³ to 224 kg/m³ with an average weight of 160 kg/m³ when dried.

## Drying

Balsa is extremely difficult to air dry from the green condition. In Ecuador, end racking is practised, and stock 100mm thick dried to below 20 per cent moisture content in 14 to 21 days. It has been reported that heavy degrade sometimes accompanies this rapid drying. In Puerto Rico, 25mm stock air dried under cover in stickered piles, dried to 17 per cent moisture content in $4\frac{1}{2}$ months with moderately heavy degrade in the form of cupping, bowing and twisting, and very slight surface checking. Kiln drying requires great care to avoid warping, splitting, case-hardening, and 'toasting' the wood. Once the wood is dry, it is stable in use ; changes in atmospheric conditions cause only minor shrinkage or swelling.

Green, freshly cut balsa generally contains from 200 to 400 per cent moisture. Soaked specimens have been recorded at 792 per cent moisture. To overcome this tendency to soak up moisture, balsa is often given water-proofing treatments with paraffin wax, water-repellents, water-repellent preservatives, varnish, or paint.

## Strength

For its weight, balsa is a strong timber, but in comparison with European redwood (*Pinus sylvestris*), balsa has about half the strength in bending and stiffness and about 70 per cent the strength in compression parallel to the grain. It is about 40 per cent weaker in bending, and 20 per cent less stiff than obeche (*Triplochiton scleroxylon*). The heartwood of balsa has only half the strength of the sapwood.

## Durability

Perishable.

## Working qualities

Very easy to work with sharp, thin-edged power or hand tools and has practically no dulling effect on cutting edges. It takes nails and screws readily but owing to the softness of the wood does not hold them well. Gluing is the most satisfactory method for fastening or holding the wood in use. It can be stained and polished but it absorbs much material used in the process.

## Uses
Balsa is used for many special purposes involving heat, sound, and vibration insulation; buoyancy, for such purposes as life-belts, floats for fishing nets and mines, buoys, water sports equipment, and hydroplane floats; lightness, for toys and models, surgical splints etc; resilience, for protective packing for delicate instruments, glass, and ceramics.
The thermal conductivity of balsa is stated to be 0.045 W/m° C per 25mm of thickness.

# BAROMALLI

*Catostemma* spp., principally          Family: Bombacaceae
*C. commune* Sandw., and *C. fragrans* Benth.

## Other names
baramanni (Guyana); flambeau rouge (French Guiana).

## Distribution
Common and locally abundant in Guyana and French Guiana. A further species, *C. sclerophyllum* Ducke is found in Brazil, but is less common.

## The tree
The trees are of excellent form, unbuttressed, with long, slender, cylindrical trunks. *C. fragrans* grows to a height of 45.0m with a diameter of 1.2m on the best sites, but is usually about 0.6m to 0.9m in diameter with clear boles 21.0m to 27.0m long. *C. commune* is smaller, and the commercial sized trees are usually 0.6m in diameter, and 18.0m to 21.0m high.

## The timber
The heartwood is dull yellowish-brown to pinkish-brown in colour, distinct, but not sharply defined from the lighter yellow-ish-brown sapwood. The grain is straight, and the texture coarse. It is a soft, light to medium-weight wood, weighing about 592 kg/m³ when dried. An unusual feature of the wood

151

is the presence of included phloem which appear as resin streaks of varying length on longitudinal surfaces, and as concentric arcs on the end grain. The rays are large, and produce a distinctive 'silver grain' figure similar to silky oak.

## Drying
Dries rather slowly without excessive degrade although some end splitting and surface checking is liable. It has a large movement tendency.

## Strength
Similar to ash in bending strength and work to maximum load, but inferior to that timber in modulus of elasticity (stiffness), hardness and shear.

## Durability
Non-durable.

## Working qualities
Works easily with only a moderate blunting effect, but chipped grain is liable due to the bands of phloem. Can be stained and polished, but a lot of filling is required. Glues and nails without difficulty.

## Uses
Dry and wet cooperage, utility plywood, light construction, cheap furniture.

# BASRALOCUS

*Dicorynia guianensis* Amsh. syn        Family : Leguminosae
*D. paraensis* Benth.

## Other names
angélique. It is frequently used locally as a substitute for teak, and for this reason is sometimes called teck de la Guyane.

## Distribution
A common tree in French Guiana and Surinam. It is said not to extend into Guyana and the lower Amazon Basin. It is particularly abundant in eastern Surinam and western French Guiana.

## The tree
A large, well-formed tree, with heavy buttresses, it attains a maximum height of 45.0m and a diameter up to 1.5m on the best sites. It is possible to obtain long lengths and large dimension stock for marine piling and similar work.

## The timber
Two distinct types of basralocus wood are recognised, ie, angélique rouge, and angélique gris, the former being darker in colour than the latter, but both have the same physical and mechanical properties. In general, the heartwood of both types is a russet colour when first cut, turning to a lustrous brown on exposure, often with a reddish cast, which is more pronounced in angélique rouge. The sapwood of this type is also more reddish. The grain is usually straight, and the texture medium; the wood weighs about 720 kg/m$^3$ when dried.

## Drying
Moderately difficult to dry because of its tendency to dry rapidly, under which conditions the wood tends to split and check. Large dimension stock should be close piled in air drying to reduce this tendency.

## Strength
When green, basralocus is similar to teak in most strength properties, but in the airdry condition it is superior to teak in all mechanical properties except tension perpendicular to the grain.

## Durability
Very durable.

## Working qualities
The wood works satisfactorily but this depends upon the silica content. It finishes smoothly and glues moderately well.

## Uses
Marine construction, bridges, decking, sleepers. In respect of resistance to marine borers, the type known as angélique gris is considered to be more resistant than angélique rouge, a factor probably brought about by the higher concentration of silica in the former type. Tests at Madison showed angélique gris to contain 1.31 per cent silica as opposed to 0.61 per cent in angélique rouge.

# 'BOXWOOD—MARACAIBO'

*Gossypiospermum praecox* P. Wils.      Family: Flacourtiaceae
Syn *Casearia praecox* Griseb.

The name boxwood, originally the English equivalent of *Buxus sempervirens*, has been extended to cover a number of botanically unrelated species with wood resembling true boxwood in general character. See also San Domingo boxwood (*Phyllostylon brasiliensis.*)

## Other names
palo blanco (Dominican Republic); 'Venezuelan', 'Colombian', 'West Indian boxwood' (UK); pau branco, castelo, zapateiro (Brazil); zapatero (Venezuela).
Note   Zapatero is also a name given to purpleheart (*Peltogyne* spp), in Trinidad, and to surudan (*Hyeronima* spp), in Panama.

## Distribution
West Indies, Venezuela and Colombia.

## The tree
A clean growing, small tree attaining a height of about 20.0m and a diameter of 150mm to 300mm.

## The timber
The wood is lemon-yellow to almost white, with little or no distinction between sapwood and heartwood. It is a hard, compact wood, with a fine, uniform texture and straight grain. It weighs about 800 kg/m³ when dried.

## Drying
It is generally shipped as small, round billets, and since the wood is not resistant to decay, it is desirable to store them in airy conditions in order to avoid discoloration. Sawing the logs down the middle, and cross-piling the half logs with the round side up, is a suitable method.

## Working qualities
Easy to carve and turn and finishes very smoothly, taking a high polish. It splits more readily than does true boxwood (*Buxus sempervirens*).

## Uses
Carving, turnery, shuttles, spindles, jeweller's burnishing wheels, knife handles, piano keys, mathematical instruments. It is sometimes dyed black to imitate ebony.

## 'BOXWOOD—SAN DOMINGO'

*Phyllostylon brasiliensis* Cap.　　　　Family : Ulmaceae

## Other names
baitoa (Dominican Republic)

## Distribution
West Indies, Mexico, South America.

## The timber
The heartwood is lemon-yellow in colour, occasionally tinged with brown and darker streaks. The sapwood is yellowish or nearly white. It has a fine, even texture, and is generally straight grained, but is sometimes wavy or curly grained. The wood is hard and heavy, weighing about 950 kg/m$^3$ when dried.

## Working qualities
It is easy to carve and turn and is capable of a high finish.

## Uses
It is not considered as of such good quality as 'Maracaibo boxwood', but is used for similar purposes, and for mallet heads.

## BRAZILWOOD

*Caesalpinia echinata* Lam.　　　　Family : Leguminosae
syn *Guilandina echinata* Spreng.

## Other names
Bahia wood, Pernambuco wood, Para wood (UK).

## Distribution
Eastern Brazil from Bahia southwards.

## The timber
The heartwood is of a fairly uniform bright orange to orange-red, which deepens to rich dark red on exposure. The sapwood is

nearly white and is sharply defined from the heartwood. The wood is hard and heavy, weighing about 1280 kg/m$^3$ when dried. It is generally straight grained with a fine, uniform texture, and has a bright, lustrous surface.

## Working qualities
It is not very difficult to work, and finishes very smoothly with a high natural polish.

## Uses
Although for many years brazilwood has been regarded as a dye-wood, it is also highly prized as a timber for violin bows because of its resilient nature. It is also used for parquet flooring, sleepers and constructional purposes. It is considered to be highly resistant to decay.

# CANAFISTULA

*Cassia ferruginea* Linn.                     Family: Leguminosae

## Other names
cassia, guarucaia.

## Distribution
Occurs in Brazil in the Zona da Mata, and is common and abundant in the Rio Grande, Minas Gerais, and Sao Paulo regions.

## General characteristics
The tree is botanically related to the laburnum (*Laburnum* spp.), and canafistula timber has some of the characteristics of laburnum being hard, heavy, and close-grained. The wood is pale reddish-brown in colour, with a medium texture and somewhat irregular grain. It weighs between 800 and 900 kg/m$^3$ when dry. The dark, concentric bands marking the season's growth, appear on longitudinal surfaces in contrast to the lighter coloured parenchyma that surrounds and links the fairly large pores.
The timber needs care in drying in order to avoid splitting and checking. Works reasonably well, and is capable of a good, smooth finish. Takes a good polish.

## Use
Construction flooring, cabinets and turnery.

# CANELA

Canela is a general trade name used, with qualifying adjectives, to describe a group of timbers mainly of the *Lauraceae* family. This is a group of aromatic trees and shrubs and the literal meaning of the word canela is cinnamon, a reference to the fact that many of the trees that produce canela contain an essential oil in the bark, often extracted commercially by distillation. The oil is used for various purposes, particularly in medicine.

The following types of canela are common to Brazil.

## *Cryptocarya mandioccana* Meiseen

### Common names
canela batalha, canela branca, canela lajeana.

### Distribution
Abundant in the Estado de Sao Paulo.

### General characteristics
The wood is light brown in colour, often with darker brown spots or patches, or it may be darkish-grey, or olive-grey. It has an interlocked or irregular grain, with a medium to coarse texture, and is moderately heavy, weighing about 720 kg/m³ when dried.

### Drying
Probably difficult to dry. From the green to 15 per cent moisture content, shrinkage values are ; 4.2 per cent radially, 9.9 per cent tangentially and 16.5 per cent volumetrically.

### Uses
Used locally for making tables, lining and skirtings, and for boxes and crates. The wood is non-durable.

## *Beilschmiedia* spp.

### Common names
canela oiti, canela branca sedosa, canela tapinha.

### Distribution
Abundant in the Estado de Sao Paulo.

## General characteristics
The colour of the wood varies from light brown to very dark brown or dark greyish-brown, sometimes with black streaks. The wood is lustrous, the grain straight or irregular, and the texture is medium to coarse. It weighs about 736 kg/m$^3$ when dried.

## Drying
Probably difficult to dry. It has a high differential shrinkage; 4.5 per cent radially, 10.5 per cent tangentially, and 16.0 per cent volumetrically.

## Uses
Used locally for interior trim, tables, and boxes. The wood is non-durable.

## *Nectandra mollis* Nees

## Common names
canela preta, canela escura, canela ferrugem, canela parda, louro preto.

## Distribution
Occurs on high ground and the slopes of the mountains of Mantiqueira and do Mar, from Esperito Santo to Santa Caterina.

## General characteristics
The wood is variable in colour, ranging from light greyish-brown to dark greyish-brown, with frequent, long, dark streaks. There is a slight superficial lustre, the grain is straight or irregular, and the texture is medium. Weight about 705 kg/m$^3$ when dried.

## Drying
Reported to dry well with little degrade. Shrinkage values, from green to 15 per cent moisture content are, 3.5 per cent radially, 7.3 per cent tangentially and 11.8 per cent volumetrically.

## Uses
Furniture, and general construction

## *Ocotea pretiosa* Nees

### Common names
canela sassafras, canela funcho.

### Distribution
Occurs from the south of Minais Gerais to the Rio Grande do Sul.

### General characteristics
The wood is very variable in colour, ranging from light yellowish-grey to light greyish-chestnut brown, often with dark longitudinal veins appearing on the wood surface. The wood has a high lustre, the grain is straight, occasionally diagonal, and the texture is medium and uniform. It weighs about 752 kg/m³ when dried.

### Drying
Its high differential shrinkage suggests it to be a difficult wood to dry. In drying from the green to 15 per cent moisture content, it shrinks 4.1 per cent radially, 11.0 per cent tangentially, and 16.0 per cent volumetrically.

### Uses
It is used in Brazil for furniture and constructional purposes, but its greatest use is probably for the extraction of essential oil which is produced by distillation not only of the bark, but also from the wood, roots and leaves.

## CANJERANA

*Cabralea cangerana* Sald.                    Family: Meliaceae

### Other names
canherana, canjarana vermelha, pau santo (Brazil).

### Distribution
Brazil and Argentine.

### The timber
A red-coloured wood which darkens on exposure to dull maroon; occasionally lighter coloured with dark streaks. The

159

texture is medium to coarse, and the grain is generally straight. The wood weighs about 705 kg/m³ when dried.

## Drying
Requires care to avoid end splitting and surface checking.

## Working qualities
Canjerana works well and finishes smoothly. There is a tendency for wavy grained stock to chip out in planing.

## Uses
Joinery, furniture, carving.

# 'CEDAR—SOUTH AMERICAN'

*Cedrela* spp. but mainly *C. fissilis* Vell.        Family: Meliaceae

## Other names
'Brazilian', 'Peruvian', 'Guyana cedar' according to origin (UK); cedro, cedro batata, cedro rosa, cedro vermelho (Brazil).

## Distribution
One or more species of *Cedrela* occurs in every country south of the United States, except Chile. In some countries where a particular species does not occur naturally, it has been widely planted and naturalized. In a commercial sense, there is no fundamental difference between the wood of the different species grown on similar sites. The growth conditions do, however, have a bearing on the type of wood produced from trees grown on different sites.

In some countries or areas, there may be two forms of a single species; one form is located along streams or on sheltered moist sites and provides light-coloured woods of relatively low density; the other form occurs on well-drained hillside forests or on drier sites, and produces denser, more deeply coloured wood.

*Cedrela* species are found particularly in Guyana. French Guiana, Surinam and Brazil, where it occurs from the Amazon to Santa Caterina and Sao Paulo, and westwards to the borders of Paraguay.

160

## The tree
A large tree attaining a height of 27.0m to 40.0m and a diameter of 0.6m to 1.25m or more. Clear boles of 15.0m to 24.0m are common above the buttresses which extend up the tree for 1.5m to 3.5m.

## The timber
The heartwood is pinkish to reddish-brown when freshly cut, becoming red or dark reddish-brown, occasionally with a purplish tinge, after exposure. The sapwood is whitish-grey or pinkish in colour. Generally speaking, cedar resembles the lighter grades of Honduras mahogany, but colour depends largely on the age and growth conditions of the tree ; the darker-coloured wood is usually produced by trees grown on the drier sites, while that from young trees, especially those of very rapid growth in the open, is less fragrant, lighter in colour, softer, but somewhat tougher than that of older trees or more slowly grown forest trees.

The texture is usually medium, although the darker-coloured woods may have a coarser texture than the lighter woods. The grain is usually straight, but occasionally is interlocked. It weighs about 480 kg/m$^3$ when dried, as an average, but the weight is governed by growth conditions.

Most wood has a characteristic odour similar to that of coniferous cedar. It is semi-ring porous or ring porous, and therefore has a visible growth pattern on tangential surfaces. Some wood shows gum streaks.

In some areas, especially in Guyana, two distinct types of wood are recognised ;

1   The darker variety, more straight grained, with a coarser texture and more volatile oil. This type of wood usually comes from trees grown on the drier sites.
2   The lighter coloured variety, not so straight grained, with a finer texture and less volatile oil.

## Drying
Generally an easy wood to dry, either in the open air or in a kiln. It dries at a rapid rate with only very minor warping, and only slight checking and end splitting. Under some conditions knots tend to split badly. There is a tendency for individual pieces to distort or collapse during kiln drying, but this may be controlled by using a low temperature schedule.

161

## Durability
Durable.

## Working qualities
Easy to work with both hand and machine tools, with very little dulling of cutting edges. It planes to a clean surface and normally finishes smoothly, but all knives and saws must be kept sharp to avoid the slight tendency to woolliness. The presence of gum in some logs gives a little trouble in planing and polishing, but, in general, the wood stains and polishes well after suitable filling. It takes nails and screws well, and can be glued satisfactorily. It can be peeled cold for veneer and plywood.

## Uses
Furniture, panelling, cabinet-making, light boat parts, canoes, clothing chests, cigar boxes, light construction, high-class joinery, plywood. In the tropics it is used for flooring and also for general house construction. Cedar has almost all the desirable properties required of a first-class wood; the one exception is the tendency for some material to exude gum that mars the wood surface.

## CEIBA

*Ceiba occidentalis* Burke                Family: Bombacaceae

## Other names
'Honduras cottonwood'. This name is confusing and should be discontinued.

## Distribution
Widely distributed throughout tropical America.

## The timber
A soft, light-weight timber, varying somewhat in colour, from greyish-white to yellowish-brown or pinkish-brown. It weighs about 350 kg/m$^3$ when dried, is not durable, and is liable to discoloration by sap stain fungus.

## Uses
It is used locally for boxes and crates, toys, interior trim and cheap carpentry.

162

# CEREJEIRA

*Amburana cearensis* (Fr. Allem.)        Family: Leguminosae
A.C.Sm.

## Other names
amburana, emburana, cumaré, cerejeira rajada. (Brazil)

## Distribution
Brazil, mainly in the eastern states, eg Valley of the Rio Doce Bahia, and Bodoquenha and Sepetuba.

## General characteristics
A uniform yellow to medium-brown coloured wood with a pinkish tint, and with a straight or irregular grain and a coarse texture. The wood is mildly scented, and weighs about 600 kg/m³ when dried.
It is reputed to be a stable wood in service, with small shrinkage similar to that of teak, but this is contradictory to published values which are as follows,
Teak        radially 1.5 per cent; tangentially 2.5 per cent.
Cerejeira    radially 2.9 per cent; tangentially 6.2 per cent.
While there is a similarity in differential shrinkage, the much higher tangential shrinkage of cerejeira suggests care is needed in drying the wood close to the final equilibrium moisture content likely to be achieved in service.
The wood is said to be fairly easy to work and finish, but care is needed to produce a smooth surface when irregular grain is present.

## Uses
Reported to be durable, the wood is used for carpentry and joinery, flooring, and vehicle bodies. It is not unlike iroko in appearance, with contrasting parenchyma, and would seem to be potentially suitable for other uses such as boat building.

# COIGUE

*Nothofagus dombeyi* Bl.        Family: Fagaceae

## Other names
coihue (Chile).

## Distribution
Occurs along the coast and up the river valleys of Chile.

## The tree
The tree generally attains a height of 21.0m to 30.0m and a diameter of about 1.0m. On favourable sites, much larger specimens can be found.

## The timber
Coigue is the beech of the southern hemisphere, and somewhat resembles European beech. The sapwood is pale pinkish, and may be very wide in some logs. The heartwood is pale brown with a distinct pinkish hue varying to reddish or yellowish-brown. The texture, grain, and density are rather variable; the wood weighs about 640 kg/m$^3$ when dried.

## Drying
Very difficult to dry without serious distortion and collapse.

## Strength
Generally inferior to European beech in all strength categories.

## Durability
Moderately durable.

## Working qualities
Works readily with hand and machine tools, but its variable density tends to cause some tearing out; tools should therefore be kept sharp.

## Uses
Vehicle bottoms, crates and boxes, joinery.

# COURBARIL

*Hymenaea* spp.                    Family: Leguminosae

## Other names
jatoba, jatai, farinheira, jatai amarelo, jatai vermelho (Brazil); locust (West Indies and Central America).

## Distribution
*Hymenaea courbaril* L., the best known of the species, occurs from southern Mexico through Central America to northern Brazil, Bolivia and Peru. In Brazil it is relatively abundant in Amazonia and Sao Paulo, and rare in Estados do Sul.

*Hymenaea davisii* Sandw., grows only in Guyana, where it is confined to the north-central and north-east districts. It occurs as an occasional tree on sandy or loamy soils within the rain or seasonal forests.

## The tree
*Hymenaea* species are generally large trees, growing to a height of 40.0m but more usually 30.0m or a little less, and a diameter of 0.6m to 1.2m. The trunk is usually free of branches for 12.0m to 24.0m, well formed and basally swollen or buttressed in large trees.

## The timber
The heartwood is salmon-red to orange-brown, becoming russet to reddish-brown when seasoned. The sapwood is usually wide, and white, grey, or pinkish, sharply demarcated from the heartwood. The grain may be straight, but is more commonly interlocked, and the texture is medium to coarse. The heartwood is frequently marked with dark streaks and sometimes shows a golden lustre. It weighs about 910 kg/m$^3$ when dried.

## Drying
Slightly difficult to air dry, with a slight tendency to warp, check and caseharden.

## Strength
Courbaril is a very strong, hard, tough wood, generally superior to oak in all properties except compression perpendicular to the grain in which both species are slightly inferior to oak.

## Durability
Moderately durable, but considered to be non-durable when a high proportion of sapwood is present.

## Working qualities
Its high density makes courbaril moderately difficult to work. It nails badly, but holds screws well. It turns satisfactorily, glues well, finishes smoothly, but does not take a high polish.

## Uses
Courbaril is used locally for general construction, looms, naves and felloes of wheels, carpentry and joinery. Its high shock resistance renders it suitable for sports goods and tool handles in place of ash; as flooring and stair treads it provides a very wear-resistant surface. It is suitable for steam-bent boat parts in place of oak, and is used for planking, tree-nails, gear cogs and wheel rims.

Second-growth timber usually has a wide sapwood, and for that reason has been recommended for veneer since it is said to work well into natural and blond-finish furniture.

# DUKALI

*Parahancornia amapa* (Huber) Duke    Family: Apocynaceae

## Other names
amaapa, mampa (Surinam); amapa (Brazil); naranja podrida (Peru).

## Distribution
Dukali occurs infrequently throughout the Amazon Basin, in Guyana, and French Guiana, and in Surinam.

## The timber
The heartwood is dull white to pale cream or pinkish, with straight grain, moderately fine texture, and is fairly lustrous. It is one of the few tropical American hardwoods suitable to some extent as a substitute for the white pine imported from the USA. The wood is moderately hard and firm, and weighs about 592 kg/m$^3$ when dried, which is a little heavier than white pine, and lighter than pitch pine.

## Drying
No information is available.

## Strength
No information is available, but its uses would suggest it to be similar to European redwood.

## Working qualities
Easy to work and finishes smoothly.

## Uses
In Guyana, dukali is used for general carpentry, interior joinery, doors and windows, concrete forms, and match-boxes. It is also recommended for plywood manufacture. It is not resistant to decay, and is liable to become badly sap-stained.

# ESPAVEL

*Anacardium excelsum* (Bert. &            Family : Anacardiaceae
Balb) Skeels

## Other names
caracoli, maranon.

## Distribution
Tropical America.

## The tree
Espavel reaches a height commonly from 18.0m to 23.0m but occasional specimens in Nicaragua have been 45.0m tall. The diameter is around 1.0m.

## The timber
The sapwood is generally very wide, occupying a large proportion of the log ; it is pale brown with a reddish cast, and can vary from 150mm to 250mm in width. The heartwood is variable in colour, ranging from pale brown with a reddish cast to yellowish-brown, often streaked with darker colour. The grain is typically interlocked, giving a pronounced stripe figure on quartered surfaces. The texture is coarse, and the wood weighs about 528 kg/m³ when dried.

## Drying
Moderately difficult to dry, it has a somewhat variable drying rate, and tends to warp and check.

## Strength
A relatively weak timber. Although only slightly lighter in weight than mahogany, it is at least 25 per cent weaker in bending, compression and shock resistance, and 15 per cent weaker in stiffness.

## Durability
Moderately durable.

## Working qualities
Rather difficult to work due to the woolly nature of the wood. There is a distinct tendency for the grain to tear out during planing and a reduction of cutting angle is necessary in order to produce a reasonable finish.

## Uses
General carpentry and joinery, boxes and crates.

# FAVEIRO

*Enterolobium* spp. Principally          Family : Leguminosae
*E. timbouva* and *E. contortisiliquum*

## Other names
timburi, tamboril, timbouva.

## Distribution
Frequent in Estado de Sao Paulo, and common in Rio Grande do Sul, Brazil.

## General characteristics
A light brown coloured wood, not unlike iroko in appearance, with a coarse, uniform texture and interlocking or irregular grain. Its weight is inclined to be variable, but averages 480 kg/m$^3$ when dry. It is reported to dry without undue degrade, and to remain stable in service.
Easy to work, and capable of a good finish. The dust is said to be irritant.

## Uses
A moderately durable timber, it is used locally for carpentry, interior joinery and furniture.

# FREIJO

*Cordia goeldiana* Hub.                    Family : Boraginaceae

**Other names**
frei jorge (Brazil) ; cordia wood, jenny wood (USA).

**Distribution**
It occurs in Brazil, mainly in the Amazon Basin.

**The tree**
Freijo attains a height of about 30.0m and a diameter of 0.6m or slightly more.

**The timber**
The wood is dark brown, not unlike teak; on quarter-sawn surfaces the rays produce an attractive figure, lighter in colour than the ground tissue. The grain is straight, the wood is rather coarse-textured, and weighs about 590 kg/m$^3$ when dried.

**Drying**
Dries readily and well, with only a slight tendency for end splits to develop.

**Strength**
Freijo has strength properties similar to those of teak, but is a tougher wood.

**Durability**
Durable.

**Working qualities**
An easy timber to work, but cutting edges must be kept sharpened to avoid grain tearing during planing and moulding. It takes stains and polish satisfactorily, but some grain filling is necessary as a rule. It glues well.

**Uses**
Panelling, joinery, furniture, boat decking as a substitute for teak, vehicle bodies, and is used in Brazil for cooperage.

# GONÇALO ALVES

*Astronium fraxinifolium* Schott        Family : Anacardiaceae
and *A. graveolens* Jacq.

## Other names
zebrawood (UK) ; tigerwood (USA).
While it has long been recognised that shipments of gonçalo alves may contain both species, Brazilian authorities nevertheless make clear distinctions between *A. fraxinifolium* and *A. graveolens.* According to Madeiras do Brasil 1965 they are described as follows,
*Astronium fraxinifolium ;* gonçalo alves, urunday-para, mura, bois de zebre.
*Astronium graveolens ;* chibatao, guarita, urunday, aderno.

## Distribution
*A. fraxinifolium* occurs southwards from Bahia, in the valley of the Rio Doce to the Minas Gerais, while *A. graveolens* extends north and east of Sao Paulo into Minas Gerais, and particularly in the State of Parana from the frontier of Paraguay to the Atlantic.

## The timber
Gonçalo alves (*A. fraxinifolium*) is reddish-brown in colour, richly mottled with darker spots and streaks. In the late eighteenth and early nineteenth centuries it was used in the UK as a substitute for rosewood, and appeared on the UK market again in the 1930's under that name. It has a medium texture, irregular and somewhat interlocking grain, often with broad contrary layers of hard and soft material. It weighs about 950 kg/m$^3$ when dried.
It is a difficult timber to dry and is inclined to warp and check. Rather difficult to work, but finishes with a high natural polish. It turns well and is highly durable.
Gonçalo alves is used for high-class furniture and cabinet-making, fancy goods, and turnery.
Chibatao or guarita (*A. graveolens*) is light reddish-brown in colour, and more plain in appearance than gonçalo alves. The texture is medium, and the grain is usually straight. It weighs about 897 kg/m$^3$ when dried.
It is used in Brazil for general construction, exterior joinery, turnery.

# GREENHEART

*Ocotea rodiaei* Mez.                      Family : Lauraceae

**Other names**
Demerara greenheart (Guyana) ; black, brown, yellow, white greenheart (Trade).

**Distribution**
Greenheart occurs principally in Guyana, but also to a limited extent in Surinam, on the upper Maratakka River, and in Venezuela on the upper Cuyuni River.

**The tree**
The tree is evergreen, usually dominant or co-dominant, and reaches a height of 30.0m or more and a diameter of about 0.6m although the maximum diameter could be 1.2m.

**The timber**
The sapwood is pale yellow or greenish, and is about 25mm to 50mm wide in young trees and about 75mm wide in mature trees. The sapwood is not clearly demarcated from the heartwood which varies in colour from yellowish-green, greenish-yellow, or light olive through golden yellow, dark olive, or yellowish-brown to very dark brown, blackish, or black. The wood from decayed or defective logs has a distinctive yellow colour. The grain varies from straight to roey, and the texture is fine and uniform, lustrous and cold to the touch.
It is exceedingly heavy, weighing about 1040 kg/m³ when dried.

**Drying**
The wood air dries very slowly, with a marked tendency to check, and for ends to split, but warping is not serious, and degrade generally is not excessive. Kiln drying is slow, with considerable degrade, particularly in thick sizes, and in general material over 25mm in thickness should be air dried prior to kilning.

**Strength**
Greenheart is one of the strongest woods in tropical America. It is hard, heavy, tough, strong and elastic, but inclined to splinter when fractured. It is only about 50 per cent heavier than English

171

oak, but 100 per cent harder, 140 per cent stronger in compression and bending, and 120 per cent stiffer in bending under gradually applied loads.

## Durability
Very durable.

## Working qualities
Greenheart is moderately difficult to work with both hand and machine tools, resisting cutting similarly to tough grades of English oak. Although it dulls cutting edges rather quickly, a fine, smooth, lustrous surface can be obtained. Because of the low cleavage resistance of the wood, cross-grained or end-grain material must be machined carefully to avoid breaking-out and splintering at the exit of the tool. The wood turns easily, and gives a good finish with wax, oil, or polish without the need of a filler, but it does not take nails well and should be pre-bored for this purpose. Gluing gives fairly good results.
The acid content of the wood is very low, calculated at 0.48 per cent of acetic acid in air dry wood, which represents a very low corrosive effect on nails, spikes and metal fastenings.

## Uses
The most important use for greenheart is in marine and ship construction. In marine construction it is used for revetments, docks, locks, fenders, braces, decking, groynes, gates, piers, piling, jetties and wharves; in ship construction, for keelsons, beams, engine bearers, planking, gangways, fenders, stern posts and sheathing for whaling ships.
Greenheart is particularly well suited for all uses where a heavy, hard, straight-grained, very strong and durable wood is required, as in heavy construction, bridges, flooring, shipping platforms, and other uses where resistance to wear is required. Other uses are fishing rods, belaying pins, billiard cue butts, mortars, and picking sticks in textile mills.

# GRUMIXAVA

*Micropholis gardnerianum*
(A.D.C.) Pierre
Family: Sapotaceae

## Other names
bacomixa, bacomixava, curubixa, gumbixava branca, gumbixava vermelha (Brazil).

172

## Distribution
Brazil, particularly abundant in the coastal region of Espirito Santo and Sao Paulo.

## The timber
A fine textured, straight grained wood, of a light pinkish-brown colour, weighing about 800 kg/m³ when dried.

## Drying
No information available. It is probably a slow drying species with drying properties similar to those of crabwood (*Carapa*). The following shrinkage data extracted from Madeiras do Brasil might prove helpful, based on drying from the green to 15 per cent moisture content.

Grumixava

| | |
|---|---|
| Radial shrinkage | 3.5 per cent |
| Tangential shrinkage | 8.1 per cent |
| Volumetric shrinkage | 12.7 per cent |

Crabwood

| | |
|---|---|
| Radial shrinkage | 4.3 per cent |
| Tangential shrinkage | 7.4 per cent |
| Volumetric shrinkage | 13.4 per cent |

## Strength
Similer in all strength categories to American mahogany (*Swietenia*).

## Durability
Non-durable.

## Working qualities
Fairly easy to work with both hand and machine tools, and is capable of a good finish.

## Uses
Interior joinery, furniture, turnery.

# GUARIUBA

*Clarisia racemosa* Ruiz & Pav.                 Family: Moraceae

**Other names**
oiticica amarela, oiti amarelo (Brazil); moral bobo, moral
comido de mono (Ecuador).

**Distribution**
Guariuba is found mainly in Ecuador, Peru, and in Brazil in the
Minais Gerais, Bahia, Espirito Santo and Zona da Mata.

**The timber**
The heartwood is bright yellow becoming yellowish-brown on
exposure, clearly demarcated from the white sapwood. It has a
medium to coarse texture, slightly lustrous and a straight, but
sometimes variable grain. It weighs between 496 and 640
$kg/m^3$ when dried.

**Drying**
No information available.

**Durability**
Probably moderately durable.

**Working qualities**
Straight grained material is said to work easily, but cutters tend
to dull quickly. It is tough and strong, and finishes smoothly.

**Uses**
Constructional purposes, shipbuilding and bridges.

# HEVEA

*Hevea braziliensis*                 Family: Euphorbiaceae

**Other names**
Para rubber tree.

**Distribution**
*Hevea braziliensis* is the source of natural rubber, and for this
reason is regarded as an exotic agricultural plantation tree. It is

found in Malaysia and Sabah, but is native to the hot, damp forests of the Amazon and Orinoco river valleys of South America.

The advent of synthetic rubber has meant the clearing of many rubber plantations, either for agriculture, or for replanting with other species such as oil palms, thus making hevea wood available to the market.

## The tree
The trees in Brazil may have reached an age of 200 years with a height of 18m to 30m and a diameter of 0.6m to 0.9m but on average the diameter is about 0.5m with a relatively short bole having a pronounced taper.

## The timber
The tree produces a light hardwood, weighing between 560 and 640 kg/m$^3$ when dry. There is little distinction between sapwood and heartwood, the wood being whitish when freshly cut, becoming light brown in colour on exposure with a pink tinge. The grain is straight, the texture moderately coarse but even, and the wood has a sour smell.

## Drying
Although the wood dries rapidly, it requires much care; it should be anti-stain dipped immediately after conversion, and should be piled under cover, with closely placed sticks. There is a tendency for the timber to warp and bow, and top weighting of the stacks is recommended, as is end coating of the boards to avoid end splitting.

## Strength
No data are available.

## Durability
Perishable.

## Working qualities
The wood is said to work easily, and to plane to a smooth finish. It tends to split in nailing.

## Uses
Suitable for pulp and plywood, provided susceptibility to stain and beetle attack is suitably controlled.

# HURA

*Hura crepitans* L.                    Family : Euphorbiaceae

## Other names
assacu (Brazil).

## Distribution
Occurs naturally in moist to wet sites from the West Indies to Central America, northern Brazil and Bolivia. It is a common tree in the Amazonia region of Brazil.

## The tree
A large, well-formed tree, which on good sites attains a diameter of 1.0m to 1.5m and a height of 27m to 40m with clear boles from 12m to 22m in length. The trees are often swollen basally or have small buttresses.

## The timber
The sapwood is yellowish-white in colour, often indistinct from the heartwood, but sometimes sharply demarcated. The heartwood is cream to light buff when freshly sawn, turning pale yellowish-brown, pale olive-grey, or dark brown on exposure, or sometimes retaining its original creamy-brown colour. The innermost heartwood of large logs is often darker in colour than the outer portion.

The grain may be straight or interlocked, and the texture medium to fine. The wood is light in weight, between 320 and 430 kg/m³ when dry, has a soft, warm feel, sometimes with a woolly surface. The wood is generally lustrous, and indistinct purplish or greenish streaks, and a slight ribbon stripe on quarter-sawn surfaces, often gives the wood an attractive figure.

## Drying
Moderately difficult to air dry. It dries fairly rapidly, which tends to prevent sap stain, but rapid drying may result in warping, at times severe. A low temperature-high humidity kiln schedule, designed for slow, careful drying is recommended.

## Strength
No detailed information available.

## Durability
Moderately durable.

## Working qualities
Works and machines readily, but where deeply interlocked grain is present, care is needed in planing and moulding to avoid chipped and torn grain. Glues easily, and takes stains and nails well.

## Uses
Plywood, veneer, furniture, joinery, general carpentry.


# IMBUIA

*Phoebe porosa* Mez.                    Family : Lauraceae

## Other names
imbuya, imbuia amarela, canela imbuia (Brazil)

## Distribution
In the Parana and Santa Caterina districts of Brazil, usually growing in association with *Araucaria*.

## The timber
The heartwood is yellowish or olive to chocolate-brown in colour, frequently variegated and sometimes figured. It is a fine to medium textured wood, and the grain may be straight, but is sometimes wavy and curly. It is moderately hard and heavy, and weighs about 660 kg/m$^3$ when dried.

## Drying
Requires care to avoid warping.

## Working qualities
It is easy to work, and finishes very smoothly, and holds its place well after manufacture. There is some tendency for the grain to pick up during planing and moulding and a reduction of cutting angle to 20° is advisable.

## Uses
In Brazil it is considered a very valuable wood for furniture, cabinet-making and high-class joinery. It is not unlike walnut when finished, and has been marketed as 'Brazilian walnut' for a number of years. It is also used in Brazil for high grade flooring, panelling and gun stocks. In Europe it is used for decorative veneer.

# IPÊ

*Tabebuia serratifolia* (Vahl)          Family : Bignoniaceae
Nicholson

## Other names
yellow poui (Trinidad) ; hakia, ironwood (Guyana) ; groenhart, wassiba (Surinam) ; pau d'arco, ipê tabaco (Brazil) ; bethabara (Caribbean area generally).
Ipê is a member of the lapacho group of the genus *Tabebuia*. The timbers in this group are noted for their great strength and durability and are characterized by the presence of a yellowish powder (lapachol compound) in the vessels. This powder has the appearance of sulphur but turns deep red in alkaline solutions. The name ipê tabaco originates from the peculiar irritating effects of the dust when inhaled during sawing or planing operations.

## Distribution
The tree is found in Trinidad, Grenada, and St Vincent of the Lesser Antilles, and on the continent from Mexico through Central America and into South America to southern Brazil, including Colombia, Bolivia, Peru, Paraguay, Venezuela, Guyana and Surinam.

## The tree
*T. serratifolia* is a canopy tree, unbuttressed or with low buttresses, about 37.0m high and a diameter of about 1.0m, although in some parts of its range diameters of about 2.0m are found. Clear cylindrical boles 15.0m to 18.0m are common. According to 'Tropical Woods 99 : 1-87', trees with trunks that will square 762mm of heartwood are available in some areas.

## The timber
The heartwood is yellowish green when first cut, turning a light to dar oklive-brown with lighter or darker streaks. The sapwood is distinct, greyish-white in colour, and 38mm to 88mm wide. The texture is fine; the grain is straight to occasionally irregular, and the lustre is low to medium. Pores in the heartwood, which appear as fine yellow dots, are filled with a yellowish powder (lapachol), and appear on longitudinal surfaces as yellow lines. The wood weighs from about 960 kg/m$^3$ to 1200 kg/m$^3$ when dried. it is cold to the touch and often appears oily; very fine ripple marks are usually present.

## Drying
Despite its relatively high density. ipê is a fairly easy wood to dry. it dries rapidly with slight warping, cupping, twisting, end and surface checking occurring. A slow drying kiln schedule is recommended.

## Strength
The wood is hard, tough and strong, and compares favourably with greenheart.

## Durability
Very durable.

## Working qualities
Ipê is moderately difficult to work especially with hand tools, and has a blunting effect on cutting edges. Ripsaws are subject to some heating when cutting thick material; fine dust escaping from the gullets packs between the saw blade and cut surfaces. This can be partially overcome by using a fairly wide tooth-pitch and increased rate of feed if practicable, or alternatively a reduced spindle speed. Using swage-set saws somewhat thicker than the standard gauge is also advisable. A good finish is usually obtained when planing or moulding plain-sawn material, but a cutting angle of at least 15° is recommended to eliminate chipping of quarter-sawn stock. The timber stains and polishes well and requires little grain-filler, but pre-boring is required before nailing to prevent splitting and bending of nails.

179

## Uses

Its strength, toughness, resilience, and very high resistance to insects and decay make it ideal for such purposes as bridge building, naval construction and dock work. It is suitable for turnery, vehicle bodies, factory flooring, decking, cabinet work and carpentry, tool handles, walking sticks and fishing rods, and archery bows. Figured logs have been cut into veneer satisfactorily.

## ITAUBA

*Mezilaurus itauba* (Meisn). Taub.          Family : Lauraceae

### Other names
itauba amarela, itauba abacate, itauba preta. (Brazil).

### Distribution
Brazil, in Amazonas, and generally in Obidos and Santarem.

### General characteristics
A medium to dark brown wood with an olive cast. It has a uniform, medium texture, and the grain may be straight, interlocked, or undulating. A fairly hard and heavy wood, it weighs about 820 kg/m$^3$ when dried.

It is difficult to dry, and has a relatively high differential shrinkage. In drying from the green to 15 per cent moisture content the radial shrinkage is 2.3 per cent, and the tangential shrinkage 6.6 per cent, with a volumetric shrinkage of 12.1 per cent.

Itauba is rather difficult to work, generally because of the interlocked grain.

### Uses
Highly resistant to decay, itauba is used locally for carpentry and joinery, naval and civil construction and other external usage. Reported to be resistant to *Teredo*.

## JACARANDA PARDO

*Machaerium villosum* Vog.          Family : Leguminosae

The name jacaranda is often confusing since it is applied to many species (some entirely unrelated). One important group of jacarandas is the product of the genus *Machaerium*.

## Other names
Jacaranda, and jacaranda with qualifying adjectives as follows, amarello, do cerrado, escuro, do mato, paulista, pedra, roxo (Brazil).

## Distribution
Brazil, in the region from the Parnaiba River to the Rio Grande, and in the Minas Gerais southward to Sao Paulo.

## The timber
The wood resembles rosewood, but is lighter in colour, being pinkish-brown to violet brown, and not so highly figured. It is a somewhat fibrous wood, with a coarse texture, and undulating grain. It is variable in weight, but averages 850 kg/m³ when dried.

## Uses
Furniture, cabinet-making.


# JATAI PEBA

*Dialium guianense*                    Family: Leguminosae

## Other names
guapaque, tamarindo, jatai mirim.

## Distribution
Various species of *Dialium* occur in the tropics; some produce keranji of Malaysia, and about ten species occur in Africa. *D. guianense* is an important species in Brazil, where it occurs in the Vale do rio Doce, Minas Gerais, Esperito Santo, and Estado do Para.

## General characteristics
The sapwood is usually wide, buff-coloured, and distinct from the heartwood which varies from golden-brown to reddish-brown in colour, with a medium, uniform texture, and generally straight grain. The wood contains resin canals and silica, and is hard, tough, and heavy, weighing about 1120 kg/m³ when dry. Care is needed in air drying to avoid deep surface checking. Although rather hard to saw, the wood works moderately well in

other operations, being fairly easy to plane. It takes and holds nails and screws well, and can be glued, stained, and polished. It is a durable species.

## Uses
A very resilient and tough timber with outstanding tensile strength properties, suitable for all types of structural use, and especially in marine situations. In tests carried out near Abidjan on the Ivory Coast, *D. guianense* was classified resistant to *Teredo* and *Bankia*, and showed a high resistance to the same borers in tests at Fort Amador on the Pacific Coast of America.

# JEQUITIBA

*Cariniana* spp.                    Family: Lecythidaceae

## Other names
jequitiba rosa (Brazil); abarco (Colombia); bacu (Venezuela).

## Distribution
Widely distributed in Brazil, and is also found in Colombia and Venezuela.

## The tree
Jequitiba grows to a height of 30.0m or more and diameters of 1.0m to 1.2m. The boles are usually clear of branches for 18.0m to 24.0m.

## The timber
The sapwood is pale greyish-brown and not clearly defined from the heartwood which varies in colour from yellowish, to pinkish or reddish-brown, sometimes marked with darker streaks. The grain is straight, and the texture medium. The weight is variable, generally from 496 to 688 kg/m$^3$ when dried.

## Drying
No information available.

## Strength
Similar to oak.

## Durability
Durable.

## Working qualities
Works easily and is comparable in this respect to American mahogany. Sharp cutters must be maintained in planing in order to avoid woolly surfaces. It glues and polishes well, but tends to split when nailed.

## Uses
General construction, joinery and cabinet-making ; it is also used in Brazil in ship and boat building in place of mahogany.

# KABUKALLI

*Goupia glabra* Aubl.                Family : Celastraceae

## Other names
cupiuba (Brazil) ; kopie (Surinam) ; goupi (Guyana).

## Distribution
Found throughout Guyana, and is widely distributed in Surinam ; it is also a common tree in the uplands of the lower Amazon and the hinterlands of Colombia.

## The tree
A large buttressed, semi deciduous tree, it grows to 39.0m and a diameter of 1.0m, but usually no more than 0.6m. Clear boles 12.0m to 24.0m are common, and short logs squaring 0.75m, and logs 18.0m long and squaring 0.3m or a little more are fairly common.

## The timber
The heartwood is orange-tan to dark russet brown, often with fine blackish streaks, and the sapwood is a light pink, brown, or yellowish colour, sometimes with yellow dots about 50mm wide. Undried wood has an unpleasant odour that dissipates during drying, but it still apparent in dried wood. The texture is medium to coarse, and the grain may be straight but more commonly is interlocked and somewhat harsh. The wood weighs about 832 kg/m³ when dried.
The Guyana Forest Department recognises two distinct varieties,

1 'White', with harsher timber, coarser texture, roey grain, and lighter in colour and weight
2 'Brown', with finer texture, smoother wood, straighter grain, and heavier in weight.

## Drying
Kabukalli is moderately difficult to dry. It dries at a moderate rate with only slight degrade in the form of warp, checking and casehardening.

## Strength
It is a hard, tough and strong wood, but for its weight it is a little below average, the greatest deficiency being in shock resistance.

## Working qualities
Moderately difficult to work; smooth surfaces are obtained in sawing, boring, and turning, but it is difficult to plane, and a cutting angle of 15° is necessary to prevent chipped and torn grain. It polishes smoothly, but usually requires a fair amount of filler. It tends to split when nailed.

## Uses
Durable, heavy construction, marine uses in non-*Teredo* waters, flooring, bridge decking, truck bodies.

## KINGWOOD

*Dalbergia cearensis* Ducke.                    Family: Leguminosae

## Other names
violete (Brazil); violetta, violet wood (USA).

## Distribution
South America, mainly from Brazil.

## The timber
The sapwood is white in colour, clearly defined from the variegated coloured heartwood which is finely striped. It is a rich violet-brown, shading sometimes to almost black, and streaked with violet-brown, black, blackish-violet, or golden yellow.

The wood is uniformly fine-textured, with a bright lustre, and is very heavy, weighing about 1200 kg/m³ when dried.

## Working qualities
Although hard and heavy, the wood is fairly easy to work, but sharp tools and cutting edges are needed to produce a very smooth finish. It is capable of a fine, waxy natural finish.

## Uses
Because of the small size of the trees, the uses are restricted to small decorative work such as inlaying, marquetry and turnery.

# KUROKAI

*Protium* spp.                               Family : Burseraceae

## Other names
porokai (Guyana) ; tingimonie (Surinam) ; incienso (Cuba) ; elimi (Brazil).

## Distribution
Although indigenous to other tropical American areas, it is more common in Guyana, where it occurs frequently in the marsh forests and in most of the rain forests. Three species are generally found in Guyana, ie *Protium crenatum* Sandw. (most abundant), *P. decandrum* March., and *P. sagotianum* March.

## The tree
Fairly tall trees, 27.0m high and a diameter of 0.5m or a little more.

## The timber
The pale buff coloured sapwood is sharply defined, and the heartwood is pinkish-brown, marked by irregularly spaced darker brown lines. The grain is straight to interlocked and irregular, and the texture is uniform and fairly coarse. The wood has a rather high lustre, and weighs about 528 to 640 kg/m³ when dried.

## Drying
Requires care ; too rapid drying can result in excessive defects developing.

185

## Strength
A moderately strong wood comparable to mahogany in most strength categories.

## Durability
Non-durable.

## Working qualities
The bark of these species contains large quantities of oleo-resin which makes conversion from the log rather difficult. This could be partially overcome by debarking prior to conversion. Dry timber saws and planes to a smooth surface with some tendency for the grain to lift, which may require considerable sanding in order to obtain a good finish. The wood polishes well but requires much grain filling.

## Uses
Masts, spars, house-framing, interior panelling, cabinet-making, and furniture, for which use it is preferred in Guyana to crabwood.

## 'LAUREL, CHILEAN'

*Laurelia sempervirens* R. & Pav.          Family: Monimiaceae
Syn *L. aromatica* Juss.

## Other names
huahuan (Chile).

## Distribution
Chile.

## The tree
A relatively small tree of the mixed forest, it occurs in nearly pure, but small stands, attaining a height of 14.0m to 15.0m and a diameter of 0.6m or a little more, with a straight bole.

## The timber
The wood is of a greenish-yellow colour, with darker streaks of brown, grey, or purple. In appearance it is similar to American whitewood, having similar grain and texture. It is a light-weight hardwood weighing about 510 kg/m³ when dried.

## Drying
Although the wood appears to dry fairly easily, it has a definite tendency to collapse, but it responds well to reconditioning treatment.

## Strength
About equal to abura.

## Durability
Non-durable.

## Working qualities
Works very easily with both hand and machine tools, but cutting edges must be kept sharp, when a good, clean finish can be obtained. Nails, screws, and glues satisfactorily, and takes stains and polishes well.

## Uses
Flooring (light domestic), and similar uses to which American whitewood is applied. The streaky colour and irregular shrinkage suggests it is not very suitable for plywood manufacture.

Note: A similar wood is also found in Chile, ie *Laurelia serrata* Ph. known as tepa, or laurela. This is a soft, not very strong wood, weighing about 480 kg/m³ when dried. It is used in Chile for boxes and crates, beehives, wood pulp, turnery and plywood.

## *LICANIA* SPECIES

Family: Rosaceae

The genus *Licania* contains at least nine species which produce very hard, heavy, and strong timber, all similar in appearance, but marketed under a variety of names, but principally as kauta, kautaballi and marish. They have a very high resistance to marine borer attack, due primarily to their high silica content. Their botanical classification, common names, and areas of distribution are as follows.

Marish is produced from *Licania buxiflora* Sandw., *L. densiflora* Kleinh., *L. micrantha* Miq., and *L. macrophylla* Benth., either as a single species or in combination.

187

Kauta is composed of three species; *L. laxiflora* Fritsch., *L. mollis* Benth., and *L. persaudii* Fanshawe and Maquire.
Kautaballi is usually made up from *L. venosa* Rusby,. and *L. majuscula* Sagot.

## Other names
marish, marishballi (Guyana) ; anaura (Brazil) ; kauston (Surinam) ; gris-gris (French Guiana).
kauta, only.
kautaballi, farsha (Guyana) ; bois galuette (French Guiana).

## Distribution
Guyana, French Guiana, Surinam, and Brazil. Kautaballi is one of the most common commercial species of Guyana, marish is fairly plentiful while kauta is relatively scarce.
*Licania macrophylla* is reported to occur frequently in the overflow woodlands of the Amazon and in the upland forests of the lower Amazon region.

## The tree
The species generally produce trees 24.0m to 30.0m high, unbuttressed, with a diameter of 0.5m. The boles are usually cylindrical, and 15.0m to 18.0m long. Marish is the largest tree of the group sometimes being 1.0m in diameter.

## The timber
The sapwood is tan in colour and usually clearly demarcated from the heartwood which is yellowish-brown to brown or dark brown, sometimes with a reddish tinge. It is straight grained, with a fine, close texture, and is very dense, hard, heavy and strong. Most species contain abundant silica. The weight of the various species vary individually, but they average as follows. Marish weighs from 930 to 1090 kg/m³ when air dry, kautaballi weighs from 1073 to 1216 kg/m³, and kauta, the heaviest of the group, weighs from 1120 to 1280 kg/m³ when dry.

## Drying
According to species, the timbers are rated easy to moderately difficult to air dry, but degrade in all species is generally light, being slight surface checking, warping and casehardening.

## Strength
There is considerable difference between the various species of *Licania* in strength properties, but as a rough guide they are comparable to those of greenheart, except in shock resistance and tension perpendicular to the grain, when *Licania* species are inferior.

## Durability
Probably moderately resistant.

## Working qualities
Owing to the hardness and high silica content of these species, they are difficult to work and cutting edges are rapidly blunted. When saws and cutters are kept sharp, the wood can be planed to a smooth surface, bored or sawn satisfactorily.

## Uses
In general, local production and use of these timbers is controlled by the fact that axe and adze are the most suitable means of working the wood. This, combined with the moderate resistance to decay, has rather restricted the use to a secondary role, for example, in Guyana, kauta is used for sleepers, mine timbers and charcoal, kautaballi is used for house framing, mine logging, shingles and charcoal, and marish for house framing and paving blocks. In Surinam, marish is used for posts, river-bank structures and sub-structures for bridges in water subject to *Teredo*.

The various species of *Licania* are primarily of interest because of their resistance to marine borers, presumably because of their high silica content. Tests at Yale University indicate this to range from 1.01 per cent in *L. mollis* to 2.82 per cent in *L. majuscula*. In marine exposure tests carried out at Wrightsville, USA, *L. macrophylla* was rated as the most highly resistant species of 37 timbers tested.

# LINGUE

*Persea lingue* Nees.                      Family : Lauraceae

## Other names
line, litchi (Chile).

## Distribution
Widely distributed in Chile between latitudes 30° and 33° S.

189

## The tree
A medium-sized tree attaining a height of about 18.0m and a diameter of 0.1m.

## The timber
The wood is rather variable in colour, being yellowish-brown to reddish-brown with a golden sheen. The grain is straight to interlocked and the texture is fine to medium. The wood weighs about 580 kg/m$^3$ when dried.

## Drying
No information available.

## Strength
Reported to be intermediate in strength properties between mahogany and oak.

## Durability
Moderately durable.

## Working qualities
Similar to the lighter grades of mahogany in working properties. Sharp cutting edges, and a reduction of cutting angle to 20° are necessary to avoid tearing of interlocked grain and to produce a good finish. It takes stains and polish satisfactorily.

## Uses
Furniture, joinery and interior construction. It is used in Chile for flooring and for gun-stocks.

## LOURO INAMUI

*Ocotea barcellensis* Mez                    Family : Lauraceae
syn. *Nectandra elaiophora*

## Distribution
Generally in the Amazonas region of Brazil.

## General characteristics
A medium brown-coloured wood, variable in weight, but averaging 670 kg/m$^3$ when dried. The grain is usually straight,

and the texture medium to slightly coarse. The wood contains an aromatic, volatile oil which imparts a spicy odour to freshly sawn wood.
It is reported to dry without difficulty, and to work well, and it has a reputation for durability, and stability in service.

**Uses**
Potentially attractive for furniture, joinery, boat building and flooring.

# LOURO PARDO

*Cordia* spp., principally *Cordia*       Family: Boraginaceae
*trichotoma* (Vell.) Arrab.

**Other names**
Louro is used in Southern Brazil for *C. trichotama* ; elsewhere in South America it usually refers to timbers of the Lauraceae family.
louro amarelo, louro da serra, cascudinho, claraiba (Brazil) ; peterebi (Argentine & Paraguay)

**Distribution**
Brazil, from the Amazon region southwards to Sao Paulo and the Rio Grande do Sul.

**General characteristics**
The wood is light greyish-yellow when freshly cut, but after seasoning and exposure assuming a pinkish cast. The grain is usually straight, and the texture is fairly coarse, and the wood has a slight lustre. It weighs about 650 kg/m$^3$ when dried.
It is reported to dry easily without excessive degrade, and to remain stable in service. It shrinks 4.3 per cent radially, 7.9 per cent tangentially, and 13.2 per cent volumetrically, when dried from the green to 15 per cent moisture content.

**Uses**
It is said to work easily and to finish well, and as an attractive light-coloured wood is potentially suitable for furniture and joinery. In Brazil it is used for joinery, chests, coffins, and for slats for venetian blinds.

191

# LOURO, RED

*Ocotea rubra* Mez.                                    Family : Lauraceae

## Other names
determa (Guyana); wane, teteroma, bewana (Surinam);
grignon rouge, grignon franc (French Guiana) ; louro vermelho
(Brazil).

## Distribution
Red louro grows in Guiana, Guyana, Surinam, and the lower
Amazon region of Brazil, and also in Trinidad. In Guyana it is of
occasional to frequent occurrence on sandy or loamy soils within
the rain and seasonal forests, while in Surinam it is more plentiful on the brown sandy soil in the Saramacca district.

## The tree
The trees are evergreen, without buttresses but usually basally
swollen, frequently 27.0m to 30.0m high and a diameter of
0.6m to 1.0m although much larger trees occasionally are
found. Their boles are cylindrical and clear of branches for
12.0m to 24.0m, but taper may be considerable, especially in
trees less than 0.6m diameter. It is possible to obtain timbers
12.0m long and squaring 0.75m of heartwood, and spars up to
21.0m to 24.0m long and 350mm in diameter at the small end.

## The timber
Although *O. rubra* is botanically related to greenheart (*O.
rodiaei*), the woods are quite different. The heartwood of red
louro is a deep salmon red when freshly cut, becoming a light
reddish-brown with a golden sheen when dried and resembling
Honduras mahogany. It also has a rather uniform colour with
pink or yellow streaks occurring occasionally. The well defined
sapwood is creamy, grey, or creamy-brown and about 25mm to
50mm wide. The grain is either straight or roey, which occasionally shows as ribbon-like bands on radial surfaces, when
quarter-sawn wood is attractively figured. The texture is coarse
and uniform, with numerous tyloses showing as shiny deposits

in the vessels.
The wood is unusually free of knots and other defects. It is moderately hard and heavy, and weighs about 640 kg/m³ when dried.

## Drying
The wood is moderately difficult to dry, largely due to the slow diffusion rate of free moisture through the wood. Thick stock tends to remain moist in the centre for a considerable time, and any attempt to speed up the drying is likely to cause collapse.

## Strength
The wood is somewhat below average in strength properties for tropical woods of similar density. In resistance to bending and compression parallel to the grain it is about equal to American white oak, but it is inferior to that timber in other mechanical properties.

## Durability
Durable.

## Working qualities
Red louro works easily and well, comparing favourably with the denser grades of mahogany in this respect. There is a need to maintain sharp cutting edges to avoid grain lifting, while the coarse texture requires the use of fillers to produce a good polished finish. Its weathering characteristics are very good, there being very little loss of surface smoothness, and remarkably little surface checking or warping, in unpainted wood exposed to the elements. It surpasses teak in its ability to resist the absorption of moisture. It glues easily and well.

## Uses
In Guyana, the wood is used for furniture, greenhouse glazing bars, interior and exterior construction, boat planking, turnery, vehicle bodies, cabinet-making, drawing boards, dowel rods, etc. Its natural resistance to moisture should qualify its use for such purposes as tight cooperage, tanks and vats, and boat parts, and it has also been recommended for plywood and veneer. Although less resistant than greenheart in respect of marine borers, it is suitable for piling and construction in waters where *Teredo* is not the main hazard.

# MACACAUBA

*Platymiscium* spp.                    Family : Leguminosae

## Other names
macacaúba preta, macacaúba vermelha, macacawood (Brazil) ; nambar (Nicaragua) ; vencola, roble colorado (Venezuela).

## Distribution
Amazon region of Brazil, and is fairly abundant in Bahia region. Occurs also in Central America.

## The timber
The sapwood is yellow, clearly demarcated from the heartwood which varies from reddish-brown to rich red or rose-red with darker veins and streaks, and a somewhat lustrous surface. The wood has a uniform, medium texture, but the grain is generally irregular and interlocked. It is a hard, heavy wood, weighing about 960 kg/m³ when dried.

## Drying
No information available.

## Working qualities
The various species of *Platymiscium* vary in their working properties from moderately difficult to difficult, but they are all generally capable of producing a smooth finish, and they polish well.

## Uses
High-class joinery, furniture, musical instruments (marimba keys), flooring, bridge decking, heavy construction. The wood is considered durable.

# MAHOGANY, BRAZILIAN

*Swietenia macrophylla* King.             Family : Meliaceae

## Other names
araputanga, cedro-i, acajou, mogno, aguano.

## Distribution
Widely distributed in Central and South America. These notes refer more specifically to the timber found in Brazil, where it

occurs over a wide area from the Vale do Sao Francisco, the Zona da Mata, and Estado de Minais Gerais. It is particularly abundant in Caceras, and near the Tocantins and Araguaia rivers.

### The tree
Often very large, attaining a height of 45m with a diameter of 2m or more above the heavy buttress.

### The timber
The wood is variable in colour, ranging from pale or medium reddish-brown to deep rich red, the darkest specimens resembling Cuban mahogany, *S. mahagoni*. The colour of the wood, its texture and grain, varies according to the locality of growth, and limited tests suggest the wood extracted from the Araguaia area more closely resembles Central American mahogany in appearance, medium texture, and shallow interlocking grain, while that from Caceras had a more marked interlocked grain, and that from the Tocantins area had similar characteristics.

It would be unwise to assume that these characteristics hold good for each locality of growth, but rather that some variation can be expected. Madeiras do Brasil, a publication of the Instituto de Pesquisas Tecnologicas, Sao Paulo, merely states the grain may be straight or interlocked, and the texture medium and uniform.

Selection of logs and timber for export purposes is the criterion on which the timber must be judged on suitable characteristics, however, and the importance of the wood should not be overlooked.

The dry weight of the wood is also variable, ranging from light, 420 kg/m$^3$ to heavy, 800 kg/m$^3$, but the Brazilian publication referred to above gives the dry weight as 640 kg/m$^3$ which suggests that in general, Brazilian mahogany is some 20 per cent heavier than Central American mahogany.

### Drying
Dries easily and well, without appreciable warping or checking. The presence of tension wood can result in longitudinal shrinkage in kiln drying.

## Strength
The strength properties are good for a timber of its weight. It compares closely with Central American mahogany in bending strength, but is slightly less stiff. It is also a harder wood than that from Belize.

## Durability
Durable.

## Working qualities
A fairly easy wood to work, either with hand or machine tools. The presence of tension wood gives rise to fuzzy surfaces, and deeply interlocked grain causes some grain tearing on quarter-sawn surfaces. The use of sharp cutting edges and suitable sanding is generally sufficient to produce a good surface. The wood takes glue, stains and polish excellently. It also produces good veneer.

## Uses
Furniture, cabinets, high-class internal and external joinery, boat building, flooring, veneer.

# MANBARKLAK

*Eschweilera* spp.                    Family : Lecythidaceae

Although some 80 species of *Eschweilera* occur from the West Indies to eastern Brazil, very few contribute to the production of manbarklak or black kakaralli.
*Eschweilera longipes* (Poit) Miers together with *E. subglandulosa* (Steud) Miers produces manbarklak (Surinam) ; black kakaralli (Guyana), and mahoe noir (French Guiana). Other names for *E. subglandulosa* are barklak, kakaralli, toledo wood, and guatecare.
Surinam is the main source of manbarklak from these two species. At least 15 species of *Eschweilera* grow in Guyana, and produce timber known as kakaralli, but a high proportion of the trees are of *Eschweilera sagotiana* Miers which produces black kakaralli which closely resembles the wood of *E. longipes* and *E. subglandulosa*. Manbarklak grows abundantly along the river estuaries in Surinam, and is a fairly frequent tree in the rain and seasonal forests of Guyana.

## The tree
The trees vary from small to very large. They are normally 27.0m to 36.0m in height, and about 0.5m in diameter, but occasionally are 1.0m or slightly more. They are somewhat fluted or slightly buttressed and have moderately well-formed boles 12.0m to 18.0m long.

## The timber
The sapwood is 38mm to about 112mm wide, creamy-tan in colour and not well defined from the heartwood which is greenish-yellow when freshly cut turning to brownish-buff on exposure, and sometimes showing black streaks. The wood is straight-grained, with a fine, uniform texture, and contains a high percentage of silica; samples of *E. longipes* show up to 2.43 per cent of silica, and *E. subglandulosa* up to 1.31 per cent based on oven-dry weight of the wood. Both these species have similar density and weigh about 1073 kg/m$^3$ when dried.

## Drying
Dries at a moderate rate with a tendency to slight end splitting and surface checking.

## Strength
Extremely hard, tough, and strong, nevertheless it is only average or below average in strength for wood of its density in the green state. It improves substantially in strength during drying, and compares favourably with greenheart in static bending, elastic resilience and shear, but is inferior to greenheart in stiffness and tension across the grain (bearing strength). Manbarklak however, is outstandingly high in shock resistance.

## Durability
Very durable.

## Working qualities
Manbarklak is difficult to work because of the high silica content. Cutting edges are quickly dulled by the silica and the wood's inherent hardness, but when knives are kept sharpened an extremely smooth surface is obtained. Rammer caps are required when driving manbarklak piling because of the ease with which it splits.

197

## Uses
It has a very good resistance to marine borers and is therefore ideal for marine and fresh water construction and piling. Good strength and high resistance to wear or abrasion makes the timber well suited for factory flooring, ice sheathing for boats, and bed plates in pulp-mill equipment.

Although manbarklak is abundant in Surinam, most commercial supplies come from Guyana, at least at the present time.

# MANDIO

*Qualea* spp.                 Family : Vochysiaceae

## Other names
mandioqueira (Brazil) ; groenfoeloe, wato-kwari (Surinam) ; grignon fou, grignon indien (French Guiana).

## Distribution
Three species of the genus *Qualea* are marketed individually or in mixture in northern South America. They are *Q. rosea* Aubl., *Q. coerula* Aubl., and *Q. albiflora* Warm. (*Q. glaberrima* Ducke). They occur in Guiana, Guyana, Surinam, and in the states of Para, Maranhao, and Amazonas in Brazil.

## The tree
The trees are generally tall, sometimes attaining a height of 60.0m with long clear boles and often of very large diameter, but usually, commercial trees are about 45.0m tall, with a well formed bole about 15.0m to 18.0m long and about 0.6m in diameter.

## The timber
The sapwood is cream-coloured, from 32mm to 56mm wide, and except for *Q. rosea* is rather clearly demarcated from the heartwood ; this is pinkish-brown when freshly cut, turning a light reddish-brown with a definite coppery or golden look. Superficially the wood resembles Spanish cedar (*Cedrela*) and quaruba or yemeri (*Vochysia*). There is parabolic figuring on the tangential face due to differences in colour density of the growth rings. The texture is medium to coarse, and the grain is straight to sometimes interlocked. All three species are moderately heavy, weighing about 608 kg/m³ when dried.

## Drying
Moderately difficult to dry, both in the open air and in the kiln. It dries rapidly but with a tendency to twist and for crook to develop, and with slight end and surface checking.

## Strength
It is similar to ash in many of its strength properties, but inferior to that wood in hardness, compression, shear and cleavage.

## Durability
Moderately durable.

## Working qualities
Silica in the wood dulls saws and cutting edges rather rapidly, and chipped or torn grain is likely to occur when roey grain is present. It glues satisfactorily, but requires care in polishing.

## Uses
Joinery, flooring, interior trim, furniture. In Holland it is used as a substitute for oak in marine piling and sheeting in non-*Teredo* waters.

# MANNIBALLI

*Inga alba* Wiild.                    Family: Leguminosae

## Other names
prokonie (Surinam) ; bois pagoda, bois sucre (French Guiana).

## Distribution
Guyana, Surinam, French Guiana, and the Rio Negro area of Brazil.

## The tree
Medium sized, and without buttresses, it grows to a height of 30.0m and a diameter of about 0.5m although larger trees are occasionally found. The bole is often fluted, varying from good to poor in form, and up to 15.0m in length.

## The timber

The whitish sapwood is not very distinct from the pale reddish-brown to reddish-white heartwood that is often streaked with darker colours. The grain is straight to roey, and the texture coarse; the wood is lustrous, and weighs about 576 kg/m³ when dried. The sapwood is liable to discoloration from sap-stain fungi.

## Drying

Moderately difficult to dry; it dries rapidly with a distinct tendency to warp in the form of cup, twist and crook.

## Strength

Tests carried out in Holland suggest it to be average or better in all strength categories for a wood of its weight.

## Durability

Non-durable.

## Working qualities

Reported to be easy to work, and finishes to a smooth surface.

## Uses

Manniballi is one of the timbers recommended by the Guyana Government for use in their housing programme for structural purposes, flooring, interior trim and partitions. Its low radial shrinkage (1.1 per cent, green to 15 per cent moisture content) suggests quarter-sawn flooring would remain remarkably stable. It is also used for plywood, furniture, light cabinet-making, boxes and crates.

## MASSARANDUBA

*Manilkara bidentata* A. DC.                Family: Sapotaceae
Syn. *Mimusops bidentata* A.DC. Chev.

## Other names

balata (Guyana); bolletrie (Surinam); nispero (Panama); maparajuba (Brazil).

## Distribution

The trees are native to the West Indies, Central America, and northern South America.

## The tree
Commonly a large, well-formed tree, reaching a height of 30.0m to 45.0m with diameters of 0.6m to 1.2m and occasionally up to 1.8m or more. Usually without buttresses, but often basally swollen.

## The timber
The heartwood is light red to rose red when freshly cut, turning dark reddish brown on exposure; sapwood whitish or pale brown, distinct, but not sharply demarcated from the heartwood. The grain is usually straight but sometimes interlocked, and the texture is fine and uniform. The wood is hard and extremely heavy, and weighs about 1050 kg/m$^3$ when dried.

## Drying
Generally reported to be difficult to dry, tending to develop severe checking, warp, and casehardening, and requiring care in piling to assure a slow rate of drying. However, reports from Puerto Rico state that 25mm lumber air dried to 19 per cent moisture content in four months with only a small amount of degrade in the form of very slight cup, crook and bow, and without apparent surface checking.

## Strength
Similar or superior to greenheart in bending strength, shock resistance, hardness, shear, and in across-the-grain properties of compression and tension, but slightly weaker than greenheart in compression parallel to the grain (crushing strength) and in elastic resilience in bending, and quite inferior in stiffness.

## Durability
Durable to very durable.

## Working qualities
Moderately easy to work despite its high density. It machines and finishes to a very smooth surface. The wood takes a fine polish and has the appearance of walnut. Gluing requires special care because of the wood's resistance to absorption of moisture.

## Uses
Steam bending, boat frames, shuttles and loom harnesses, billiard cues; heavy construction, bridges, flooring in industrial plants, stair treads.

# MORA

*Mora excelsa* Benth                 Family: Leguminosae

### Other names
prakue (Guyana); peto, witte mora (Surinam); mahot rouge (French Guiana).

### Distribution
*M. excelsa* is widely distributed in Guyana, Surinam, and French Guiana, and less so in Trinidad and Venezuela.

### The tree
It is an outstanding tree, normally 30.0m or more high and about 1.0m in diameter, with clear boles for 15.0m to 18.0m above the huge buttresses. Logs are commonly 9.0m to 12.0m long and 0.5m to 0.75m butt diameter, but some run to 18.0m long and 1.5m through the butt.

### The timber
The sapwood is 50mm to 150mm wide, yellowish in colour, and distinct from the heartwood which is dark brown, reddish brown, or dark red, streaked with white or brown lines. The grain is straight or more commonly interlocked, very variable, and often has an attractive bird's eye, wavy, or ribbon-grain figure. The texture is usually coarse, giving a rather harsh feel to the wood.
Two types are recognised, ie black and white; black mora has more heartwood, and is heavier and more durable than white mora. The average weight is about 1040 kg/m³ when dried.

### Drying
Rather difficult to dry. It dries slowly and degrade in the form of surface checking and distortion is liable to be appreciable unless care is taken.

### Strength
A very strong, stiff, hard, tough wood with strength comparing favourably with oak, but generally a little higher in most categories.

## The timber
The wood is very similar to manni (*Symphonia globulifera*), being dark yellowish-brown or orange brown with conspicuous greyish parenchyma markings. The sapwood is yellowish white which is reported to gradually turn blackish on exposure. The grain is usually straight but may occasionally be irregular, while the texture is medium. The wood is moderately hard and heavy, and weighs on average 800 kg/m³ when dried.

## Drying
No information is available.

## Strength
An elastic wood, which however is inferior to oak in practically all strength categories.

## Durability
Durable.

## Working qualities
Works fairly well, but takes nails poorly. It polishes very well.

## Uses
It is used locally for piling, boat building, furniture, turnery, carriage-building and general construction. In French Guiana it is also used for rum barrels, packing cases and crates. It has been suggested as being suitable for utility plywood.

## PAU AMARELO

*Euxylophora paraensis* Huber.　　　　Family: Rutaceae

## Other names
pau cetim, sateen wood.

## Distribution
Abundant in the Estado do Para region of Brazil, especially along the Tocantins river.

## General characteristics
The wood is a bright lemon-yellow when freshly cut, turning a rich golden-yellow on exposure, with a medium, uniform

texture, and a grain that is generally straight, but is sometimes wavy. It is a very handsome wood with a slight superficial lustre, weighing about 850 kg/m³ when dry.
The wood dries without difficulty, is easy to work, finishing well and taking stains and polish satisfactorily.

## Uses
A valuable wood for furniture, decorative cabinet work, flooring etc.

Note: The Portuguese, amarelo, means yellow, and this word is often applied to Brazilian woods of a yellowish or yellow colour.
Vinhatico (*Plathymenia reticulata*) is a case in point, sometimes also being called pau amarello or pau amarelo.
Pau amarelo is also a local name for fustic (*Chlorophora tinctoria*), and for arapoca (*Raputia magnifica*).

## PAU MARFIM

*Balfourodendron riedelianum* Engl.          Family: Rutaceae

## Other names
moroti (Brazil and Argentina) ; quatamba (Brazil).

## Distribution
Southern Brazil, in the region of Rio Paranapanema, Sao Paulo, and Rio do Sul. Also in Argentina.

## The timber
The wood is whitish-yellow to lemon-coloured, with little distinction between sapwood and heartwood. The texture is very fine and uniform, the grain straight or irregular, occasionally interlocked, and the wood has a medium lustre. It weighs about 800 kg/m³ when dried.

## Working qualities
Easy to work, but with some tendency to blunt the cutting edges of tools fairly rapidly. Turns well and finishes smoothly.

## Uses
A compact, fine textured wood, suitable for turnery, shoe-lasts, textile rollers and for flooring. A good substitute for maple.
Note : The name pau marfim is also used for other fine textured, pale coloured woods, especially for some species of *Aspido-sperma*.

## PAU MULATO

*Apuleia leiocarpa* (Vog) Macbride.     Family : Leguminosae
syn *A. praecox* Mart.

### Other names
garapa, grapiapunha, pau cetim* (Brazil).
*Pau cetim is often used locally to describe this wood, but it is misleading since it is also applied to pau amarelo, *Euxylophora paraensis*. There is some similarity in appearance, but the timbers are unrelated.

### Distribution
Occurs from Corrientes and Misiones in Argentina, throughout most of Brazil, to Venezuela to eastern Peru.

### The tree
In the Argentine and southern and north-eastern Brazil, the trees are usually less than 24m tall and 0.9m in diameter at maturity, but on good sites, they occasionally reach a height of 30m and a diameter of 1.2m with a clear bole of 15m to 18m.

### The timber
The sapwood is narrow and whitish in colour, while the heart-wood is mainly yellowish, varying from yellowish-brown to pinkish-yellow, tending to acquire a reddish or coppery hue after exposure. The wood is lustrous, with a straight to roey grain and fine, uniform texture. The wood is hard, heavy, tough, and strong, and weighs from 800 to 960 kg/m³ when dry.

### Working qualities
Said to be easy to work, finishing smoothly, and to be durable.

### Uses
Constructional purposes. The timber has been used for a long time locally for making canoes for use in rapids.

# PEROBA ROSA

*Aspidosperma* spp., principally             Family : Apocynaceae
*A. peroba* Fr. All., *A. polyneuron*
Muell. Arg.

## Other names
red peroba (UK).

## Distribution
Brazil, principally from the south-east regions of Goias, Minais Gerais and Sao Paulo.

## The tree
The tree is variable in height and girth according to locality but on average it reaches a height of 27.0m with a diameter of about 0.75m.

## The timber
The sapwood is white or yellowish, merging gradually into the heartwood which is very variable in colour, ranging from yellow to rose-red often with purplish or dark streaks and patches. It is hard and moderately heavy, weighing about 770 kg/m$^3$ when dried, and it is fine textured with a straight to very irregular grain. There is no superficial lustre.

## Drying
Requires care in drying to avoid distortion and splitting.

## Strength
Although heavier and harder than English oak, peroba rosa has about the same stiffness. Owing to its irregular grain tendencies, its general strength properties vary considerably, but selected material is about 30 per cent better in bending strength than oak, and 1 per cent superior in resistance to shock loads.

## Durability
Durable.

## Working qualities
Moderately easy to work with only minor blunting of cutting edges, but wild grain is liable to tear out during planing and

moulding, and a cutting angle of 20° is recommended. It glues, stains and polishes well.

## Uses
Furniture, joinery, flooring, sleepers, and sliced decorative veneer.

# PIQUIA

*Caryocar villosum* Pers.                    Family: Cariocaraceae

## Other names
pequia, piquia bravo, vinagreira.

## Distribution
Brazil, in the Amazon region and Para.

## General characteristics
The wood is straw-yellow in colour, with a pinkish tinge, rather coarse-textured, and with a grain that is either interlocked or diagonal. It weighs about 816 kg/m³ when dry.
Requires care in air drying since the wood has a distinct tendency to warp and twist. It is a moderately difficult wood to work due to variations in its texture, some zones of wood appearing less dense than others and tending to tear in sawing and planing. The wood has a smell of vinegar when worked, hence the local name, vinagreira.

## Uses
A very strong, tough, and durable wood, it is valued in Brazil for frames, knees, and floor timbers for boats, and for hubs and felloes of wheels. It is a particularly valuable timber for marine construction for piling, sea defences, etc. and is reported to be resistant to attack by *Teredo*.

# PURPLEHEART

*Peltogyne* spp.                    Family: Leguminosae

## Other names
amaranth (USA) ; koroboreli, saka, sakavalli (Guyana)'. purperhart (Surinam) ; pau roxo, amarante, (Brazil).

About 20 species of *Peltogyne* occur in tropical America, but only three are of economic importance in the Caribbean area. These are *P. pubescens* Benth., *P. porphyrocardia* Griseb., and *P. venosa* (Vahl) Benth. var. *densiflora* (Spruce) Amsh. *P. venosa* (Vahl) Benth., growing in Guyana, the Guianas, and Amazonas, Brazil, is of general economic importance in the Amazon region.

## Distribution
The commercial species of purpleheart are distributed widely throughout tropical America. Their combined range extends from Mexico through Central America and into South America to southern Brazil. There appears to be little need for discrimination between the species since all the timbers have about the same general appearance and structure regardless of their source.

## The tree
Purpleheart of Guyana is a semi-deciduous tree with small plank-type buttresses about 1.0m high in *P. pubescens*, and 1.8m to 3.6m high and sometimes spread out over a 4.5m diameter in *P. venosa* var *densiflora*. The trees reach a height of 50.0m with straight, cylindrical boles 18.0m to 27.0m high above the buttresses and with considerable taper. The diameter is generally 1.0m or slightly more.

## The timber
The sapwood is creamy white to light pinkish cinnamon streaked with light brown, from 50mm to 100mm wide, and sharply demarcated from the heartwood which is most often recognised for its unusual colouring. It is greyish-purple when freshly cut, later becoming a violet purple to deep purple through an oxidation process. In due time, the purplish colour is lost and the wood turns an attractive dark brown; this is only superficial, and by removing only a thin layer the original colouring is restored until oxidation occurs again.
The uniform texture of purpleheart varies from fine to medium. The grain is usually straight and seldom interlocked, but is sufficiently irregular, along with variation in lustre and colour to give the wood a pleasing stripe figure on quarter-sawn surfaces.
The weight of purpleheart is variable, but averages 880 kg/m³ when dried.

## Drying
Moderately difficult to dry in the open air and in the kiln. It air dries slowly with light to moderate degrade, usually in the form of slight end and surface checking and slight casehardening. Thin timber kiln dries readily, but thick material requires special care to overcome the difficulty of removing moisture from the centre of the heavy pieces.

## Strength
Purpleheart is intermediate between American white oak and greenheart in strength properties, but is outstanding in its ability to withstand suddenly applied loads and difficult strains.

## Durability
Very durable.

## Working qualities
Moderately difficult to work; the wood resists cutting and dulls cutting edges. It also exudes a gummy resin when heated by dull tools; the resin clings to cutter teeth and other tool parts and complicates machining operations. Material is best run slowly through machines equipped with high speed steel knives. A cutting angle of 15° is required to properly machine material with interlocked grain.
The wood turns smoothly and requires little sanding to bring out a good finish. It glues well, and takes stain and wax polish easily, but its purple colour is dissipated by spirit polishes; it is reported that a lacquer finish holds the purple colouring. The timber should be pre-bored for nailing to avoid splitting.
Purpleheart is knife-cut, but more generally sawn for veneers without preparatory steaming. It is considered inadvisable to steam the wood as the colouring matter known as phonicoin is partially soluble under prolonged steaming.

## Uses
Purpleheart is a relatively expensive wood of high quality, and its uses are best diverted into two kinds, (1), those requiring great strength, particularly in relation to sudden shocks and difficult strains, and (2), those requiring wood of unusual beauty or colouring. Type 1. gymnasium apparatus, diving boards, skis, mill rollers, shafts and tool handles. Purpleheart is considered Brazil's best timber for spokes in cart wheels, and

because of its dimensional stability it is used in the tropics for window frames and sliding shades. Though expensive, it is highly suitable for structural purposes, bridge building, fresh water piling, and house building, flooring, and cladding.
It is used in Guyana for knees, transoms, deadwoods, stems, and interior work in boat building.
Type 2. The unique shades and peculiar variegated or mottle colour effects make purpleheart well adapted for use in turnery, marquetry, furniture, cabinet-making, counter tops, carving, inlaying, panelling, billiard cue butts, billiard tables, flooring, and other similar uses.

# QUARUBA

*Vochysia* spp.                                    Family : Vochysiaceae

The name quaruba covers several different species of *Vochysia* of the Brazilian Amazon, and includes *Vochysia guianensis* Aubl., *Vochysia tomentosa* D.C., and *Vochysia tetraphylla* (G. F. Mey) D.C.

## Other names
quaruba jasmirana (Brazil) ; iteballi (Guyana) ; kwarie (Surinam).

## Distribution
Although quaruba is found in Guyana and Surinam, it is more abundant in Brazil, where it occurs in the lower Amazon on silt soils adjacent to waterways and on sandy soils. It makes its best development along the coastal plain.

## The tree
The trees are unbuttressed, medium to fairly large in size and range from good to poor in form. The usual mature tree is about 27.0m to 38.0m high and clear of limbs for about 21.0m with a diameter of 1.0m and sometimes much more.

## The timber
The heartwood is distinct but not always sharply defined from the greyish-pink or buff sapwood which is 38mm to 75mm

wide, and the heartwood is light pinkish-brown with a golden lustre. The texture is rather coarse, and the grain is generally straight; the wood weighs about 500 kg/m³ when dried. It is similar to the allied yemeri, but more variable in all respects.

## Drying
It is moderately difficult to dry both in the open air and in the kiln. The timber dries at a moderate to slow rate, with a strong tendency to warp and surface check. The various species making up quaruba undergo considerable shrinkage during drying, and there is a high ratio of tangential to radial shrinkage, in the order of 8.2 to 10.8 per cent tangentially to 3.9 to 4.8 per cent radially in test drying from the green to oven dry. There would appear to be every advantage in using quarter-sawn stock.

## Strength
A moderately light and soft hardwood, it is more comparable to European redwood than to oak where by comparison it is generally inferior particularly in resistance to splitting and hardness on side grain.

## Durability
Moderately durable.

## Working qualities
Quaruba works readily, but has rather poor machining properties. In sawing a fairly long tooth pitch is necessary for best results. Raised grain is a common defect on planing and moulding operations and sharp cutters are necessary to alleviate the problem, but a satisfactory finish can be obtained by sanding. The wood is not suited for turnery because of its open texture, which also necessitates a considerable amount of filling before polishing. It polishes to a good finish but water stains should not be used since they cause considerable grain raising. Quaruba takes glue, paint and nails well.

## Uses
General carpentry, boxes and crates, inexpensive furniture, and interior joinery. It is reported as suitable for plywood manufacture.

# RAULI

*Nothofagus procera* Oerst.                    Family : Fagaceae

*Nothofagus* produces the beech of the Southern Hemisphere ; *Fagus* produces the true beech of the Northern Hemisphere, eg English beech.

**Other names**
'Chilean beech', 'S. American beech'.

**Distribution**
Chile.

**The tree**
Rauli occurs in pure stands on rich soil, and attains a height of 40.0m with a clear bole of some 18.0m and a diameter of 0.75m. It was formerly very abundant, but extensive felling has depleted many stands.

**The timber**
The heartwood is a uniform reddish-brown, and the texture is fine and uniform. It is similar to mild grades of European beech, but is lighter in weight, about 560 kg/m$^3$ when dried (European beech 720 kg/m$^3$), and lacks the prominent fleck, produced by the rays on radial surfaces. There is also a tendency for rauli to show a pore ring on the tangential face.

**Drying**
Dries slowly, but with little degrade.

**Strength**
Generally inferior to European beech in all strength categories.

**Durability**
Probably durable.

**Working qualities**
Easy to work with both hand and machine tools ; finishes cleanly, and takes nails, screws, and glue well, and can be stained and polished satisfactorily.

## Uses
As a substitute for European beech, except where strength is an exacting requirement. It is suitable for domestic flooring, and is used in Chile for doors, window frames and furniture. It has been used in the UK for jigs and pattern-making in place of Honduras mahogany, and has proved satisfactory for those purposes, and also for framing for vehicles.

# ROSEWOOD, BRAZILIAN

*Dalbergia nigra* Fr. All. principally          Family: Leguminosae

## Other names
jacaranda, jacaranda da Bahia, jacaranda preto (Brazil); Rio rosewood, Bahia rosewood (UK); palissander, palissandre du Brazil (France).

The commercial name jacaranda should not be confused with the botanical name *Jacaranda copaia*, which produces futi or parapara of Brazil.

## Distribution
Occurs in Bahia, Esperito Santo, and is of most frequent occurrence in the Zona da Mata and Minas Gerais.

## The timber
The trees produce very varying examples of colour and figure in the wood which varies from chocolate or violet-brown, to a rich purple black. The timber has a somewhat oily appearance and the sapwood is nearly white in colour and sharply defined. When worked, the wood has a distinct mild, fragrant odour. The grain is usually straight, and the texture medium, sometimes gritty. It is a hard, heavy wood, weighing about 870 kg/m$^3$ when dry. It air dries slowly, with a tendency to check.

## Working qualities
It is somewhat difficult to work because of its hardness, and tends to blunt cutting edges rather quickly. It is capable of an exceedingly smooth surface but is sometimes too oily to obtain a high polish.

## Uses
Brazilian rosewood has been familiar as a decorative wood for some 200 years, both in the solid and as veneer. It is mainly used today in the form of veneer for furniture, cabinet-making, pianos, and in the solid for fancy woodware, knife handles, spirit-levels and similar.

Note: *D. spruceana* Bth. produces jacaranda do Para, also sold as jacaranda or rosewood. This occurs in Brazil in the Manaus region, Rio Parintins, Santarem and Territorio do Amapa.
It is slightly heavier and rather less decorative than *D. nigra* and is used locally for interior finish, cutlery handles, fancy woodware and furniture.

## SAJO

*Campnosperma panamensis* Standl.      Family: Anacardiaceae

## Distribution
Widely distributed from Panama to Peru, from the hillsides of East Panama and around the Santiago—Cayapas river system and into Ecuador and Colombia.

## The Tree
The tree reaches a height of 20m with an umbrella-like crown. The diameter is about 0.6m and the bole is straight and cylindrical, and free of branches for more than half the height.

## The Timber
The wood bears a superficial resemblance to that of Philippine lauan; there is little difference by colour between sapwood and heartwood, the wood varying from greyish-pink or mauve-grey to reddish-brown with a purple tinge. The growth rings are poorly defined though sometimes indicated by colour variations. The grain is interlocked, and the texture medium to fine. The wood weighs about 430 kg/m$^3$ when dry.

## Drying
Dries rather slowly. Tests at Madison suggest the same kiln schedules recommended for sande should be satisfactory.

## Strength
No information.

## Durability
Non-durable.

## Working Qualities
The wood works and machines with moderate ease. It tends to tear in planing and moulding and a cutting angle of 15° is said to give good results. It can be stained and polished, and takes glue well.

## Uses
Light construction, interior joinery, boxes and cases, etc.

# SANDE

*Brosimum utile* (HBK) Pittier.                    Family: Moraceae

## Other Names
cocal.

## Distribution
Ranges from Costa Rica to Pacific Ecuador. Considerable volumes of standing timber occur in the forests of Pacific Ecuador and Colombia the source of practically all present shipments of sande.
*Brosimum* spp. are generally considered in two groups, ie alicastrum and utile. The first named receives its name from *B. alicastrum*, the best known of the species in their group while *B. utile* lends its name to the utile group, the timbers of which are appreciably lighter in weight than those of the alicastrum group.

## The Tree
*Brosimum utile* is a small medium-sized tree about 20m to 24m tall, with a diameter of 0.6m.

## The Timber
There is no distiction by colour between sapwood and heart-wood, the wood being a uniform yellowish white to yellowish-brown or light pinkish-brown, with a straight to widely and

shallowly interlocked grain, and medium texture. It is not unlike white seraya from Sabah. Sande weighs about 528 kg/m³ when dry, as opposed to the timber of the alicastrum group, which averages 800 kg/m³.

## Drying
Early shipments to America contained a degree of tension wood, and in evaluation tests at Madison it was thought expedient to produce two kilning schedules, particularly since severe kiln conditions caused severe end checking despite the ends being coated with filled varnish. Under these conditions, some honeycombing occurred, and there was considerable darkening of the wood surface. A moderate schedule was found satisfactory for normal wood, and a low temperature schedule was produced for parcels containing excessive tension wood, as follows. (p.78)

## Working Qualities
Normal wood of sande machines easily, takes stains and polish readily, and presents no gluing problems. Where tension wood is present, as has been the case in some early shipments, some rapid dulling and burning of saws may be experienced, due to the pinching effect when stresses are released in the wood. This has resulted in a claim that the wood contains silica; chemical tests have revealed however that the maximum silica content is only 0.01 per cent of the oven dry weight of the wood, and cannot be held responsible for machining difficulties.
Fuzzy grain due to tension wood presents some difficulty in planning and moulding, and a reduction of cutting angle to 15° may be helpful. Normal wood finishes very well from the machines.

## Uses
The unfavourable reports on early shipments of sande, generally in regard to colour variations and degree of tension wood, and recommendations from America that trees with a definite lean or an obvious eccentric bole are best left in the forest, has led to more suitable selection of saw logs. Large quantities of sande are currently being shipped to Canada and the USA principally for use for door jambs.
Samples of the wood sent to the UK have been of a light brown or pinkish-brown with reasonable uniformity, and Colombian

shippers advise that 95 per cent of exportable material falls within this range.

The wood appears very suitable for interior joinery and fitments, light construction, and probably for furniture.

### Moderate kiln schedule for normal 25mm sande

| Moisture content % | Dry bulb Temperature °F | Dry bulb Temperature °C | Wet bulb Temperature °F | Wet bulb Temperature °C | EMC % |
|---|---|---|---|---|---|
| Above 60 | 130 | 54.5 | 125 | 52 | 16 |
| 60 | 130 | 54.5 | 122 | 50 | 13.5 |
| 40 | 130 | 54.5 | 118 | 48 | 11 |
| 30 | 140 | 60 | 100 | 38 | 4 |
| 18 | 180 | 82 | 130 | 54.5 | 3 |
| First equalizing | 180 | 82 | 145 | 63 | 5 |
| Final equalizing | 170 | 77 | 135 | 57 | 5 |
| Conditioning (6 hr.) | 180 | 82 | 170 | 77 | 11 |

### Low temperture kiln schedule for 25mm sande with excessive tension wood

| Moisture content % | Dry bulb temperature °F | Dry bulb temperature °C | Wet bulb temperature °F | Wet bulb temperature °C | EMC % |
|---|---|---|---|---|---|
| Above 60 | 110 | 43.5 | 105 | 40.5 | 16 |
| 60 | 110 | 43.5 | 102 | 39 | 13.5 |
| 35 | 110 | 43.5 | 98 | 36.5 | 11 |
| 30 | 120 | 49 | 85 | 29.5 | 4.5 |
| 15 | 160 | 71 | 110 | 43.5 | 3 |
| Equalizing | 170 | 77 | 135 | 57 | 5 |
| Conditioning (6 hr.) | 180 | 82 | 170 | 77 | 11 |

Reference: Kiln/drying selected Colombian Woods; McMillen and Boone:
Forest Products Journal vol. 24. No 4. April 1974.

**Strength**
No information

**Durability**
Non-durable

## SANTA MARIA

*Calophyllum brasiliense* Camb.                    Family: Guttiferae
var *rekoi* Standl.

**Other names**
guanandi, jacareuba, pau de Maria (Brazil); koerahara, (Surinam); krassa (Nicaragua).

**Distribution**
Santa Maria grows throughout the West Indies and on the continent from southern Mexico southward through Central America and into northern South America.

**The tree**
A large tree attaining a height of 30.0m to 45.0m and diameters from 1.0m to 2.0m with long clear boles up to 15.0m to 21.0m.

**The timber**
The heartwood of Santa Maria varies from pink or yellowish-pink to brick-red or rich reddish-brown, marked by fine darker red parenchyma striping. The sapwood is lighter in colour, but not always distinct, and is 38mm to 62mm wide. The texture is medium and fairly uniform, and the grain is sometimes straight but generally interlocked. The wood is attractive and usually has a ribbon figure on quarter-sawn surfaces. It is sometimes mistaken for mahogany, being somewhat similar in colour, but is heavier, stronger, and more durable in some uses. It weighs about 610 kg/m$^3$ when dried.

## Drying
Moderately difficult to air dry; it dries slowly with considerable warping and splitting; moisture is often difficult to extract from the centre of thick stock. When quarter-sawn, and dried carefully in a kiln, the material is of first-class quality.

## Strength
For its density, Santa Maria is one of the stronger woods in its class. It exceeds mahogany in all strength properties and is more comparable to American white oak.

## Durability
Very durable.

## Working qualities
Moderately easy to work, but the thin, soft parenchyma tissue may pick up badly when planing plain-sawn surfaces. This can be alleviated by reducing the cutting angle to 20°. Brown gum streaks, when present can cause rapid dulling of cutting edges. It nails, screws, glues, stains and polishes satisfactorily.

## Uses
General construction, bridge building, shingles, flooring, shipbuilding, interior construction and furniture. Produces a fair veneer.

# SATINÉ

*Brosimum paraense* Hub.                  Family: Moraceae

## Other names
muirapiranga (Brazil); bois satiné (France); satiné rubané (France).

## Distribution
Tropical America.

## The timber
The sapwood is thick and clearly demarcated from the heartwood, which varies from greyish-red to rich strawberry-red, with a golden sheen. The texture is fine, and the grain generally

straight, but somewhat variable. The wood weighs about 1010 kg/m³ when dried.

## Working qualities
Although hard, the wood is not difficult to work, glues well, and takes a high natural finish.

## Uses
Furniture, marquetry, turnery and veneer. Generally of limited availability.

# SIMARUBA

*Simaruba amara* Aubl.                Family : Simarubiaceae

## Other names
marupa (Brazil) ; simarupa (Guianas). The names aceituno and negrito are also used in Central America.

## Distribution
*Simaruba amara* is found in northern South America from Venezuela and the Guianas to Brazil and in Trinidad and Tobago. It is an occasional tree in the rain and savannah forest in Surinam ; in Guyana it is of frequent occurrence in the seasonal forests and as an occasional tree in the rain forests.

## The tree
A large, unbuttressed tree attaining heights of 42.0m and diameters of 0.6m and occasionally 1.0m. The trees have straight, cylindrical, strongly tapered boles, frequently 21.0m to 27.0m long.

## The timber
The heartwood is whitish or cream coloured when freshly cut, becoming a uniform cream colour with occasional oily streaks. There is no distinction between heartwood and sapwood, and the wood is entirely without figure except for a few, widely spaced narrow vessel lines. It has a light lustre, the texture is uniform and medium, and the grain is straight. The wood weighs about 450 kg/m³ when dried.

## Drying
The wood air dries at a rapid rate with only slight end checking, but precautions are necessary to prevent sap-staining. It has low shrinkage values, but is somewhat unstable in response to atmospheric changes.

## Strength
Simaruba is not a strong wood, being somewhat brittle and inclined to split along the grain. It is slightly inferior to American yellow poplar in bending strength, but superior to that timber in proportional limit stresses in static bending and compression parallel to the grain, hardness and shear.

## Durability
Moderately durable.

## Working qualities
Works easily and machines to a smooth, clean surface. Sharp tools and high speeds are essential in turning. Easy to paint, stain and polish but care must be taken in gluing to maintain the full strength of the wood.

## Uses
Boxes, crates, interior joinery, shoe heels, drawer lining.

# SNAKEWOOD

*Piratinera guianensis* Aubl.　　　　Family : Moraceae
and allied species

## Other names
letterwood (UK) ; amourette (France) ; bourra courra (Guyana) ; letterhout (Surinam) ; palo de oro (Venezuela).

## Distribution
The several species of *Piratinera* occur throughout the Amazon region of Brazil and extend northwards through Guyana, Venezuela, Colombia, Panama, and into southern Mexico and the West Indies. The principal species *P. guianensis* grows in Guyana, French Guiana, Surinam, Brazil, Bolivia and Trinidad.

## The tree
Snakewood is a small tree, seldom growing more than 24.0m high, and 0.3m to 0.6m in diameter. It is unbuttressed but

generally swollen at the base; the bole is cylindrical and clear of branches for 12.0m to 15.0m. It is not found in abundance at any point in its extensive range; it is a rare or occasional tree in Guyana, and is found occasionally in the rain and marsh forests of Surinam.

## The timber
Snakewood acquires its name from the peculiar markings or irregular dark spots, which resemble the letters of the alphabet, the spots of a leopard, or the skins of highly coloured snakes. The tree is slow in forming heartwood, which is the only part used commercially. A tree of 380mm diameter may, on occasions, have only 25mm to 100mm of heartwood, while a 508mm tree will ordinarily have not more than 178mm. The amount of heartwood is reported to vary with the location. Certain areas produce timber with considerable heartwood, while in other areas even large trees may not be worth felling.
The timber is exported with all the sapwood removed in the form of small logs or sticks measuring 100mm to 250mm in diameter and generally 2.0m long or slightly more. The logs are sold by weight and are reputed to be one of the most expensive woods in the trade.
The sapwood which the trade rarely sees, is light yellow to nearly white, and not clearly demarcated from the heartwood which is dark red to reddish-brown with conspicuous irregular black radial markings (speckles), or with black vertical stripes sometimes found with the speckles. The wood is extremely hard and heavy, weighing about 1300 kg/m³ when air dry. The dark areas are the result of variations in the gummy deposits that fill all the cell cavities.

## Working qualities
Snakewood is worked with considerable difficulty; it is difficult to cut, yet finishes smoothly and takes a beautiful polish.

## Uses
Walking sticks, fancy articles, drum sticks, fishing rod butts, fancy handles for cutlery and umbrellas, violin bows and archers' bows, although in this respect, while it is a traditional wood for native bows, they are not generally considered suitable for tournament or contest shooting.
Guyana, French Guiana, and Surinam, presently supply limited quantities of snakewood.

# STERCULIA

*Sterculia* spp.                               Family: Sterculiaceae

The genus *Sterculia* comprises about 60 species of trees and shrubs distributed throughout tropical and sub-tropical regions in both hemispheres. The following are a few of those species of economic importance in the American tropics.

*Sterculia caribaea* R. Br. occurs in the West Indies and is dealt with in the companion booklet 'Timbers of Central America and the Caribbean'.
*Sterculia pruriens* (Aubl.) K. Schum. is reported as a timber of importance in Guyana and northern Brazil.
*Sterculia rugosa* R. Br. is similarly reported in Guyana, and may occur elsewhere.
The following description applies to both *S. pruriens* and *S. rugosa*.

## Other names
yahu (Guyana) ; karst, castano (Honduras) ; kobehe (Surinam).

## Distribution
Both species are of occasional to frequent occurrence in the heavy forest of Guyana, where they are generally distributed throughout the near interior and Rupununi district, becoming more scattered in the eastern district.

## The tree
It is a large, unbuttressed tree, commonly 30.0m in height, and about 0.6m in diameter at maturity. The boles are cylindrical, 18.0m or more long and slightly tapered.

## The timber
A rather plain wood of fairly good quality, with a soft, light greyish coloured heartwood with numerous brown ray flecks imparting an overall brownish effect. The sapwood is not very distinct from the heartwood, and is usually about 50mm wide, and is subject to discoloration from sap-stain fungi. The grain is usually straight, and the texture medium to coarse. The wood is not especially fibrous as compared to some other members of the genus and is fairly lustrous. It weighs about 592 kg/m³ when dried.

## Drying
Sterculia is moderately difficult to dry. It tends to warp and twist, and care must be taken to slow down the drying rate.

## Strength
It is somewhat weaker in cleavage (splitting), but somewhat harder than the average for other timbers of similar density. In general it is not a wood of great strength, and should not be used where high strength is required.

## Durability
Non-durable.

## Working qualities
The wood works easily with both hand and machine tools and does not blunt cutting edges. Planed surfaces tend to be fibrous, but a reduction of cutting angle to 20° usually alleviates the problem. It nails well, and a good finish can be obtained by sanding; it takes stains and polish quite well, but a fair amount of grain filling is generally necessary.

## Uses
Light construction, interior joinery, cheap panelling, boxes and crates.

# SUCUPIRA

*Bowdichia nitida* Benth.                Family: Leguminosae

## Other names
sapupira (Brazil); black sucupira (UK).
*Bowdichia nitida* generally supplies the sucupira exported from Brazil, but the name is also applied to other species, for example to *Diplotropis racemosa*; sapupira parda, sapupira, sapupira da mata (Brazil); *Ferreirea spectabilis*; sucupira amarela, sucupirana (Brazil). (This has also been marketed as yellow sucupira).

## Distribution
Brazil, where it is abundant in northern and eastern parts of the country, especially in Para, Amazonas, Belem, and the region around Manaus.

## The tree
Sucupira is commonly 27.0m to 30.0m in height and about 0.6m in diameter, but trees up to 1.0m are occasionally found. The trunk is usually straight, cylindrical, without buttresses, and up to 18.0m in length.

## The timber
The heartwood is a dull brown to reddish-brown colour with light yellow parenchyma markings, especially noticeable on plain-sawn surfaces. The sapwood is whitish, narrow, and sharply demarcated. The texture is coarse, and the grain wavy and irregular; weight about 990 kg/m³ when dried. It is a hard, heavy, tough and strong timber.

## Drying
Rather difficult to dry. Considerable checking and cupping can occur in kiln drying unless a slow drying schedule is employed.

## Strength
No information.

## Durability
Reported to be very durable.

## Working qualities
The wood is moderately difficult to work. It saws rather easily, but is somewhat difficult to plane because of its frequent roey grain, which when present requires considerable sanding. It turns well, has good screw-holding power, and takes wax or polish satisfactorily if a filler is used.

## Uses
Furniture, turnery, and lighty duty flooring.

Note: 'Commercial Timbers of the Caribbean' refers to tatabu, a product of *Diplotropis purpurea* (Rich) Amsh. This timber is very similar to sucupira, and occurs in the rain forests in the upland areas in Guyana, Surinam, and Brazil. It is reported to be an occasional tree in Guyana, where it is marketed as tatabu, rare in Surinam, where it is known as zwarte kabbes, and fairly common in some areas of Brazil, where it is known as supupira or sucupira.

227

A reference is made to J. H. Hughes; 'Handbook of Natural Resources of British Guiana 1947'; "The qualities and properties of this wood (tatabu) have not been fully appreciated in the past. Extra work spent on its manufacture is amply repaid by the results achieved. This wood will definitely find a niche among our more beautiful woods in the future when it may be possible to exploit the forests to a greater extent than in the past". The timber is used in Guyana for heavy construction, boat building, house framing, furniture and turnery, and for flooring, and for civil and naval construction in Brazil.

## SURADAN

*Hieronyma* spp.                    Family: Euphorbiaceae

About 25 species of *Hieronyma* grow in tropical America, but only two species are of widespread importance. These are *H. alchorneoides* Fr. All. and *H. laxiflora* (Tul) Muell.-Arg.

**Other names**
suradanni, pilon (Guyana); sorodon, anoniwana (Surinam); zapatero (Panama); nancito (Nicaragua); margoncalo, uru-curana, (Brazil).

**Distribution**
*H. alchorneoides* occurs from British Honduras through Central America to Guyana, French Guiana and Brazil. *H. laxiflora* is reported in Guyana, French Guiana, Peru, Colombia and other parts of the Amazon Basin.

**The tree**
In general, the trees are large, straight, evergreen trees with spreading, rounded buttresses. They may reach 40.0m in height and a diameter of 1.0m or more, but commonly they are about 30.0m high, with a diameter of 0.6m. Clear stems may be up to 21.0m long.

**The timber**
The sapwood is pink and about 38mm wide, and the heart-wood is a reddish-brown to chocolate brown or a dark red, not unlike black walnut in appearance. Growth rings are marked by

changes in colour at the margins of each season's growth, resulting in a series of parabolic markings on the tangential face. The wood is moderately coarse in texture, and the grain is interlocked, giving a striped or ribbon-grain appearance to the wood. It sometimes contains stones of calcium oxalate. The wood weighs about 800 kg/m³ when dried.

## Drying
Moderately difficult to air dry, with a moderate rate of degrade in the form of crook, and surface checking. Slow drying is recommended.

## Strength
It is reported to have normal strength properties for wood of its density, but is rather low in shock resistance, hardness, and cleavage.

## Working qualities
It is good in all working properties except planing, the roey grain causing irregular chipping out. The wood glues well, and finishes smoothly, but scraping is sometimes necessary. A perfect finish usually requires a lot of grain filling because of the rather large pores.

## Uses
It is highly resistant to decay and marine borers, and is well suited for marine work and piling, bridges, dockwork, general construction, and is recommended for panelling and high-class interior work and furniture.

# TATAJUBA

*Bagassa guianensis* Aubl. and          Family: Moraceae
*B. tiliaefolia* (Desv.) R.Ben.

## Other names
bagasse (Guyana) ; gele bagasse (Surinam).
Note: tatajuba is also used in Brazil for fustic (*Chlorophora tinctoria*) and for species of *Clarisia*.

## Distribution
*B. guianensis* and *B. tiliaefolia* occur infrequently in Guyana and French Guiana as scattered trees in the low upland forests ;

*B. tiliaefolia* is fairly numerous in Guyana in the near interior and Rupunumi district. *B. guianensis* occurs infrequently in the Amazon Region of Brazil, where it is marketed as tatajuba.

## The tree
The tree reaches a height of 27.0m to 30.0m and a diameter of 0.6m although taller and larger specimens are found. The bole is cylindrical and 18.0m to 21.0m high.

## The timber
The sapwood is pale yellow to yellowish-white in colour, narrow, and sharply demarcated from the heartwood which is yellow with darker streaks when first cut, becoming lustrous golden-brown to russet after seasoning and exposure. The grain is medium to moderately coarse and usually interlocked, presenting a rather broad striped figure on the radial surface. The timber is hard and heavy and weighs about 800 kg/m$^3$ when dried.

## Drying
Dries slowly, but with very little degrade. It has an exceptionally low volumetric shrinkage (10.2 per cent) when drying from the green to oven dry. Tangential shrinkage is 6.6 per cent, and 5.2 per cent radially.

## Strength
Tatajuba compares favourably with oak except in shock resistance and shear.

## Durability
Durable.

## Working qualities
Easy to work and takes a high lustrous finish, and holds its place well after manufacture. It lends itself well to natural bends for boat building.

## Uses
Heavy construction, both civil and marine. Furniture, cabinets, decking and framing in boat building.

# TINEO

*Weinmannia trichosperma* Cav.                    Family: Cunoniaceae

## Other names
teneo (Chile); saisai (Venezuela); tarco (Argentine).

## The timber
The heartwood is brownish to a light reddish-brown, the lighter coloured sapwood merging into the heartwood. It has a fine uniform texture, with variable grain, and weighs from 592 kg/m³ to 705 kg/m³ when dried.

## Working qualities
Easily worked, it may be considered similar to Canadian birch.

## Uses
A general utility wood, suitable for joinery when selected, otherwise, boxes and crates, dowels, brushes, spools and bobbins.


# TULIPWOOD, BRAZILIAN

*Dalbergia frutescens* Britton var                 Family: Leguminosae
*tomentosa* Standl.

## Other names
pau rosa, jacaranda rosa, pau de fuso (Brazil); pinkwood (USA). Not to be confused with the wood of the tulip tree (*Liriodendron tulipifera*).

## Distribution
Brazil, mainly in the north-east and around Bahia.

## The timber
An attractive wood, violet-red in colour, streaked with deeper red and salmon-coloured stripes, but losing its bright colour on exposure. It has a medium to fine texture, but the grain is usually irregular. The growth is dense and hard, but the wood is liable to split after being sawn. It weighs about 960 kg/m³ when dried.

**Working qualities**
The wood is usually shipped in the form of small logs or billets, and is usually fairly wasteful in conversion. It is not easy to work, but is capable of a high natural finish.

**Uses**
Turnery, brush backs, marquetry, fancy woodware and marimba keys.

# ULMO

*Eucryphia cordifolia* Cav.     Family: Eucryphiaceae

**Other names**
gnulgu, muermo (Chile).

**Distribution**
Found mainly in the southern part of Chile.

**The timber**
The heartwood is dark reddish or greyish-brown, sometimes variegated, with a fine, uniform texture and generally straight-grained. It weighs about 630 kg/m³ when dried. It is strong, moderately hard, dries reasonably well without checking, and is not very durable.

**Working qualities**
The wood works very easily with both hand and machine tools.

**Uses**
Flooring, furniture, vehicle bodies, oars, joinery.

# VERAWOOD

*Bulnesia arborea* Engl.     Family: Zygophylacea

**Other names**
'Maracaibo lignum vitae' (UK).

**Distribution**
South America, principally from Venezuela.

## The timber
The timber of *Bulnesia* to some extent resembles that of the closely related 'true' lignum vitae (*Guaiacum* spp.). It is however, much more variable in colour and characteristics, and although sometimes used as a substitute for lignum vitae, it is unsuited for the most exacting uses to which the latter wood is applied. The colour of the heartwood gives rise to the following local descriptions, vera aceituno, v. amarillo, v. azul, and blanca, or olive, yellow, blue, or white, respectively. It weighs on average 1140 kg/m$^3$ when dried.

The main objections made to verawood are that it does not wear well, but this must be taken as a reference to its comparison with *Guaiacum* for such uses as bushings for propeller-shafts. It would seem that the biggest objection is the propensity for verawood to develop ring and cup shakes in the log.

In Venezuela the wood is much valued for its high durability and is used there for posts, and for brush backs, mallet heads and other small articles.

## VINHATICO

*Plathymenia reticulata* Benth.     Family: Leguminosae

## Other names
vinhatico castanho (Brazil).

## Distribution
Brazil.

## The timber
Although this name is applied to several genera which have similar coloured wood, the above species has been the main supplier of vinhatico received in the UK. The heartwood is a lustrous yellow or orange-brown darkening on exposure to a rich brown shade and is frequently striped with lighter and darker shades. It is medium to slightly coarse textured, straight to roey grained, and weighs from 560 kg/m$^3$ to 640 kg/m$^3$ when dried.

## Working qualities
The wood is very easy to work both with hand and machine tools. It is reported as being easy to dry and stands well

when manufactured. It is said to be a little difficult to polish, and requires a grain filler.

## Uses
Furniture, joinery, flooring, shipbuilding, shop-fitting and shoe heels. The denser specimens are satisfactory for vehicle bodies.

# VIROLA, HEAVY

*Virola* spp., but principally        Family : Myristicaceae
*V. bicuhyba*
### Other names
bicuiba becuva, bucuvucu, bicuiba branca, bicuiba vermelha (Brazil).

### Distribution
Several species of the genus *Virola* occur in Brazil, but *V. bicuhyba* which is darker in colour and heavier than the other species making up light virola, baboen and banak. It is most abundant in the Estado de Sao Paulo region, principally in the valley of the Rio Ribeiro.

### General characteristics
The heartwood varies in colour from light brown to rose-red, to more or less dark brown. It has a straight grain, the texture is medium, and the wood weighs about 670 kg/m$^3$ when dried. It has a lower shrinkage ratio than other *Virola* species, and in drying from the green to 15 per cent moisture content, it shrinks 5.3 per cent radially, 8.4 per cent tangentially, and 15.3 per cent volumetrically. Heavy virola is moderately durable, reputed to work and finish easily and well, and is popular in Brazil for furniture, carpentry, and construction. It would seem to be a stable wood in service since it is also used for rulers and for slats for venetian blinds.

# VIROLA, LIGHT

*Virola* spp., principally        Family : Myristicaceae
*V. surinamensis* (Rol) Warb.

The various species of the genus *Virola* are very similar in appearance, and to some extent in physical properties. Mostly, they resemble the lighter-coloured mahoganies.

*V. surinamensis* is the principal species producing light virola, otherwise known as baboen ordalli (Guyana) ; ucuuba (Brazil). *V. sebifera* Aubl. is the principal species producing virola, also known as St Jean rouge (French Guiana) ; banak or moonba (Surinam), and is sometimes also sold as ucuuba.
*V. melinonii* (R. Ben) A. C. Smith. produces the hoogland baboen of Surinam, but locally this description often includes a mixture of *V. melinonii* and *V. sebifera*. They are also sometimes mixed in small quantities with *V. surinamensis* and sold as virola.
*V. sebifera* and *V. melinonii* are less preferred in the plywood industry of South America generally.
*V. bicuhyba* produces commercial heavy virola which is dealt with elsewhere in this publication.
*V. koschnyi* produces banak. This timber is found more particularly in Central America and the West Indies, and is therefore dealt with in the companion publication, 'Timbers of Central America and the Caribbean'.

### Distribution
*V. surinamensis* grows in Guyana, French Guiana, Surinam, Venezuela, and the Amazon region of northern Brazil. It is also common in the West Indies.
*V. sebifera* ranges from Nicaragua through northern South America as far as Peru, Bolivia, and southern Brazil.
*V. melinonii* is found mostly in northern South America.

### The tree
Light virola is very common in the swamp and marsh forests of Surinam and is a frequent to locally common tree throughout Guyana in the riparian, mora, and marsh forests. It is abundant in the islands of the Amazon estuary of Brazil.
The tree is heavily buttressed, of medium to large size, attaining a height of 42.0m and a diameter of 1.5m under favourable conditions, but usually is much shorter, with a cylindrical bole, clear of branches to between 18.0m and 24.0m and about 1.0m in diameter.

### The timber
The heartwood is a cream or tan colour when freshly cut, darkening on exposure to a pinkish, golden brown, or deep reddish-brown, not very clearly demarcated from the lighter

coloured sapwood. The wood has a low lustre and medium to coarse texture; the grain is straight, and the wood weighs from 430 to 580 kg/m$^3$ when dried.

## Drying
Light virola has a high shrinkage ratio by comparison with Honduras mahogany. In drying from the green to oven dry condition, light virola shrinks 5.3 per cent radially, 12.4 per cent tangentially, and 17.6 per cent volumetrically, against corresponding values of 3.5, 4.8, and 7.7 per cent respectively for the mahogany. The wood accordingly has a strong tendency to cup and split radially, and for thick stock to retain its moisture despite rapid surface drying. However, provided care is taken, the wood should dry with a minimum of degrade.

## Strength
In the air dry condition, light virola is superior to Honduras mahogany in stiffness and shock resistance, but is inferior to that timber in other strength properties, being closer to those of American whitewood (*Liriodendron*) in bending strength and gradually applied loads, but inferior to that species in crushing strength and in shock loading.

## Durability
Perishable. The wood is very liable to severe fungal staining and should be anti-stain treated when used in the form of lumber.

## Working qualities
Light virola works easily and very satisfactorily. It nails, screws and glues well, and can be stained without difficulty to resemble mahogany fairly well, it gives satisfactory results in polishing and varnishing. When properly dried it holds its place well when manufactured, with practically no tendency to warp or check.

## Uses
The light virola species are used mostly for plywood particularly in Surinam, but as lumber they are suitable for many uses requiring a light, easily worked, non-durable timber. In Surinam, light virola is used for cigar boxes, coffins, crates, matches, and match-boxes. It is also used for heavy-duty battery separators, but is not considered suitable for car batteries because of

the mechanical failure of the thin sheets. Although non-resistant to decay when in contact with the ground, light virola is said to have good lasting properties below water level, and accordingly it is used in Surinam for foundation boarding below water level. In Guyana the wood is used, together with other *Virola* species, for concrete forms, general carpentry, coffins, slack cooperage, and as a backing veneer.

# WALLABA

*Eperua* spp.                                     Family : Leguminosae

## Other names
*Eperua falcata* Aubl., produces soft wallaba, while *E. grandiflora* (Aubl.) Benth., *E. jenmanii* Oliver, and *E. schomburgkiana* Benth. produce ituri wallaba. These names are commonly used in Guyana, but the following names are also in common use ; waapa, bijlhout, (Surinam) ; apa, aipe (Brazil) ; palo machete (Venezuela).

## Distribution
Mainly found in Guyana, French Guiana, and Surinam, but extending into Venezuela and the Amazon region of northern Brazil.

## The tree
Although the trees may reach 27.0m to 30.0m in height and a diameter of 0.6m or more, they grow rapidly, and are often defective due to the development of heart rot; consequently merchantable round logs are usually only about 6.0m to 7.5m long and about 0.5m in diameter. Transmission poles may be produced up to 18.0m long and with butt diameters of 200mm to 300mm.

## The timber
The sapwood is grey or greyish-white in colour, sharply demarcated from the deep red to reddish-purple heartwood. The surface of the wood is streaked with resin, which appears in concentric rings on the ends of logs with gummy exudations. The wood is straight grained, and the texture is medium to coarse. All four species mentioned are similar in appearance and weight, which is about 910 kg/m$^3$ when dried.

237

## Drying
Wallaba dries slowly with a marked tendency to warp and split. An initial period of air drying prior to kiln drying is essential if degrade is to be kept to a minimum.

## Strength
Hard, heavy, stiff and strong, it is similar to karri in most strength categories, except in resistance to shock loads and splitting.

## Durability
Very durable.

## Working qualities
Although hard, it works easily with hand and machine tools except that gum collects on the cutting edges of tools, particularly saw teeth. This can be overcome by using saws with 88mm pitch and generous gullet space. It planes to a smooth surface, but tends to char when bored. There is a tendency for the finish to be spoiled later by gum exudation, but in tests, it would seem that treatment with a grain filler and wax polish satisfactorily held back the resin.

## Uses
Wallaba is exceptionally well suited for use as transmission poles, flag-poles and posts, provided the sapwood is removed. It is also satisfactory for wharf and bridge construction, but is not considered resistant to marine borers. It is also suitable for all types of flooring and decking.

## WHITE PEROBA

*Paratecoma peroba* Kuhlm.          Family: Bignoniaceae

## Other names
peroba de campos, ipé peroba, peroba amarella, peroba branca, ipé claro (Brazil).

## Distribution
Brazil, where it is found in Minais Gerais, Esperito Santo, and the Zona da Mata, mainly in the coastal forests.

## The tree
The tree reaches a height of some 39.0m and a diameter of up to 1.5m with a straight, symmetrical bole, free from branches up to about 28.0m.

## The timber
The sapwood is white or yellowish, and clearly defined from the heartwood which is variable in colour, basically light olive brown; it may have yellow, green, or red shading. The texture is medium and uniform, and the grain is commonly interlocked, often producing a narrow stripe figure on radial surfaces. The wood is variable in weight, averaging 770 kg/m$^3$ when dried.

## Drying
Dries readily and well, although variable grain may lead to some twisting.

## Strength
Generally similar to teak, but superior to that timber in hardness and shear.

## Durability
Very durable.

## Working qualities
Generally works well and finishes smoothly, but interlocked grain may tend to pick up in planing.

## Uses
Civil and naval construction, joinery, flooring, furniture, vehicle bodies, boat building for decking.

# PART II SOFTWOODS

## ALERCE

*Fitzroya cupressoides*                    Family : Cupressaceae
F. M. Johnston

### Distribution
The tree occurs in central and southern Chile.

### The tree
Alerce is found in the low swamp forests where it grows to a height of 30.0m or more, and a diameter averaging 1.25m.

### The timber
The heartwood is brownish-red, not unlike western red cedar, but usually a little darker in colour, harder, and heavier. The weight is somewhat variable, but averages about 510 kg/m³ when dried. The grain is straight, the texture very fine and even, and in general the wood is singularly free from objectionable defects.

### Strength
Alerce is not a strong wood ; its hardness and shear strength are comparable to European redwood, but all other strength properties are much lower and similar to those of obeche (*Triplochiton*).

### Durability
Durable.

### Working qualities
Easy to work, with little blunting effect on cutting edges. The wood planes to a very good finish if cutters are kept sharpened. It glues, stains, and polishes satisfactorily.

### Uses
The high quality of alerce, which can be attributed to its very slow, uniform rate of growth, renders it ideal for many purposes where high durability and pleasing appearance are the main requirements. It is used in Chile for furniture, joinery, cooperage, masts and spars for boats.

# 'CHILE PINE'

*Araucaria araucana* K. Koch.          Family : Araucariaceae

## Other names
'Chilean pine' (UK) ; pilon, pehuen (Chile).

## Distribution
Southern Chile.

## The tree
Chile pine or monkey puzzle tree, grows to a height of 36.0m or more, and a diameter of 1.5m.

## The timber
Pale brown in colour, very similar in all respects to Parana pine, but lacking the bright red streaks common to that timber, and showing small brown flecks to a much greater degree than those appearing in Parana pine. Weight about 550 kg/m$^3$ when dried. The grain is usually straight, and the texture is even and uniform.

## Drying
Dries rapidly with a slight tendency to distort, split, and check. There is also a tendency for the wood to give up its moisture unevenly.

## Strength
No information is available.

## Durability
Non-durable.

## Working qualities
Works fairly easily, but with some slight blunting effect on cutting edges, and a tendency for the grain to chip out in planing, and to crumble when worked across the grain. It takes paint and varnish well, but stains unevenly. It can be glued satisfactorily.

## Uses
Joinery, carpentry, light construction and flooring.

# MANIO

*Podocarpus* spp., and            Family : Podocarpaceae
*Saxegothaea conspicua* Lindl.

## Other names
*Podocarpus nubigenus* Lindl.; maniu, manilihuan, maniu hembra (Chile).
*Podocarpus salignus* D. Don. (Syn. *P. chilinus* Rich.) ; maniu de la frontera (Chile) ;
*Saxegothaea conspicua* Lindl. ; maniu macho (Chile).

## Distribution
The two species of *Podocarpus* occur in Chile on swampy land and along banks of the River Maule, in a narrow belt stretching from San Ferndo to Chiloe Island. *Saxegothaea conspicua* is also native to Chile, and is found a little farther south in the swampy parts of the rain forests of Rio Mauleto.

## The tree
Commercial manio may contain timber from *Podocarpus* and *Saxegothaea*, but their wood is very similar in appearance. There are certain differences however, in the form of the individual species which, to some extent have a bearing on the drying behaviour and working properties of commercial parcels of manio.
*P. nubigenus* attains a height of 9.0m to 24.0m and a diameter of 0.75m or a little more, with a bole which is generally of good shape. *P. salignus* is not so tall, on the best sites reaching a height of 18.0m and a diameter of about 0.45m often with a distorted or twisted bole. *Saxegothaea* is similar in height and girth, and the bole is often twisted.

## The timber
The wood is pale yellow to yellowish-brown in colour, often with brown streaks, and lacking the conspicuous late-wood growth markings which characterize most of the softwoods of the northern hemisphere. The grain is usually straight, and the wood is soft, with a uniform fine texture. The weight is about 560 kg/m³ when dried.

242

## Drying
No information is available, but it is probable the drying properties are similar to those of East African podo (*Podocarpus*
spp.), which dries rapidly, but with a pronounced tendency to
distort, split and check.

## Strength
Comparable to European redwood (*Pinus sylvestris*) in most
strength categories, but harder and slightly less resilient.

## Durability
Probably durable.

## Working qualities
Easy to work, but some chipping out of the grain is liable in
planing and moulding, and in boring and mortising, and cutters
should be kept well sharpened. Manio stains and polishes
satisfactorily, but tends to split when nailed.

## Uses
Joinery, flooring, cooperage.

## 'PARANA PINE'

*Araucaria angustifolia* O. Ktze.        Family : Araucariaceae

## Other names
'Brazilian pine' (USA).

## Distribution
It occurs in Brazil in the region of Parana, Rio do Sul, Santa
Catarina, and to a lesser extent in western parts of Sao Paulo.

## The tree
The trees are straight and clean almost to their full height which
reaches a maximum of 36.0m on the best sites, but may be
shorter. The diameter is about 1.0m.

## The timber
The timber is pale brown, with a central core of darker brown
coloured wood ; it may be streaked with red, but this is some-

times absent. The grain is mostly straight, and the texture close and uniform, with very inconspicuous growth rings. It contains little resin, and varies from light and soft, to moderately hard and heavy, but averages about 550 kg/m$^3$ in weight when dried.

## Drying
In general it is more difficult to dry from the green state than most softwoods. It is variable in its drying properties and dark-coloured material dries slowly with a pronounced tendency to split and distort. Inherent stresses in the timber are liable to cause distortion when the timber is machined. Most of the drying problems occur however in the initial drying from the green and imported stock does not pose too many problems when this is kiln dried. Care must be taken however, and the inherent stresses must not be ignored; stress checks, and suitable conditioning at the end of a kiln run should ensure uniformly dried material.

## Strength
The timber is similar to European redwood in general strength properties, but it lacks toughness, which restricts its use for certain purposes.

## Durability
Non-durable.

## Working qualities
Easy to work, and planes and moulds to a clean, smooth finish. Stains, polishes, paints and glues well.

## Uses
Interior joinery, including doors. Its lack of toughness renders it unsuitable for scaffold boards and long ladder stringers. 'Parana pine' is used for plywood manufacture in South America.

# USE GUIDE FOR SOUTH AMERICAN TIMBERS

## AGRICULTURAL IMPLEMENTS

ipê, kabukalli, mandio, purpleheart, Santa-Maria, suradan

## BOAT AND SHIP CONSTRUCTION

**Decking**
cerejeira
freijo (sub for teak)
greenheart
mahogany, Brazilian
purpleheart
tatajuba
white peroba

**Framing**
greenheart
ipê
louro, red
massaranduba
mora
pakuri
piquia
purpleheart
Santa Maria
suradan

**Keels
and
Stems**
greenheart
ipê
jatai peba
kabukalli
kauta
kautaballi
louro, red
manbarklak
marish
massaranduba
piquia
purpleheart
Santa Maria
suradan
tatajuba

**Oars**
ulmo

**Planking**
cedar
courbaril
greenheart
jequitiba
louro, red
mahogany, Brazilian
santa-maria
tatajuba

**Superstructures**
andiroba
'cedar'
jequitiba
mahogany, Brazilian
purpleheart
Santa Maria

# BOXES AND CRATES

baromalli
canela batalha
canela oiti
ceiba
coigue
dukali
espavel
'laurel, Chilean'

light virola
manniballi
pakuri
quaruba
simaruba
sterculia
tepa
tineo

# CONSTRUCTION

**Heavy**
acapu
andiroba
angelim pedra
brazilwood
canafistula
greenheart
ipê
itauba
jatai peba
kabukalli
kauta
kautaballi
louro, red

macacaúba
manbarklak
mandio
marish
massaranduba
pau mulato
pakuri
sajo
Santa Maria
suradan
tatajuba
wallaba
white peroba

**Light**
baguacu
baromalli
canela preta
canela sassafras
ceiba
'Chile pine'
faveiro

gmelina
light virola
lingue
manniballi
olivillo
quaruba
sande
sterculia

# DOORS

andiroba
arariba
'cedar'

dukali
heavy virola
rauli

246

# FANCY GOODS

gonçalo alves
kingwood
louro pardo

rosewood
snakewood
tulipwood

# FLOORING

acapu
andiroba
arariba
brazilwood
canafistula
cerejeira
'Chile pine'
courbaril
gmelina
greenheart
imbuia
ipê
kabukalli
'laurel, Chilean'
lingue
louro inamui
macacaúba

mahogany, Brazilian
mandio
manio
manniballi
massaranduba
mora
pau amarelo
pau marfim
peroba rosa
purpleheart
rauli
sande
Santa Maria
sucupira
ulmo
vinhatico
white peroba

# FURNITURE AND CABINETS

andiroba
canafistula
canela preta
canjerana
'cedar'
faveiro
freijo
gmelina
gonçalo alves
grumixava
heavy virola
hura
imbuia
jacaranda pardo
kurokai
light virola

lingue
louro inamui
louro, red
mahogany, Brazilian
mandio
manniballi
muiratinga
pau amarelo
peroba rosa
purpleheart
rauli
rosewood
sande
satiné
sucupira
vinhatico

## GUN STOCKS
imbuia, lingue

## INSULATION
balsa

## JOINERY

**High-class**
alerce
andiroba
canjerana
'cedar'
cerejeira
courbaril
freijo
grumixava
hura
imbuia
jequitiba
kurokai

light virola
lingue
louro inamui
louro, red
mahogany, Brazilian
mandio
peroba rosa
sande
suradan
ulmo
vinhatico
white peroba

**Utility**
baguacu
baromalli
canela batalha
canela oiti
ceiba
'Chile pine'
coigue
dukali
espavel
faveiro
gmelina

heavy virola
hura
'laurel, Chilean'
manio
muiratinga
'Parana pine'
quaruba
rauli
simaruba
sterculia
tineo

## MARINE PILING AND CONSTRUCTION

**Under Water**
**(a)** *Teredo* **infested waters**
acapu
angelim pedra
basralocus
greenheart
jatai peba
kauta

kautaballi
manbarklak
marish
piquia
suradan
tatajuba

**(b)** Non-*Teredo* waters, in addition to above,

| | |
|---|---|
| ipê | massaranduba |
| itauba | mora |
| kabukalli | pakuri |
| louro, red | wallaba |

## Above Water

**(a)** Docks, wharves, bridges, etc.

| | |
|---|---|
| acapu | manbarklak |
| andiroba | mandio (sub for oak) |
| basralocus | marish |
| greenheart | massaranduba |
| guariuba | mora |
| ipê | pakuri |
| itauba | suradan |
| kauta | tatajuba |
| kautaballi | wallaba |
| louro, red | white peroba |

**(b)** Decking

| | |
|---|---|
| acapu | mandio (sub for oak) |
| basralocus | mora |
| brazilwood | piquia |
| greenheart | purpleheart |
| ipê | tatajuba |
| kabukalli | wallaba |
| macacaúba | white peroba |
| manbarklak | |

## TERMITE RESISTANCE (HEARTWOOD) *

**Very resistant**

| | |
|---|---|
| courbaril | massaranduba |
| greenheart | mora |
| ipé | purpleheart |
| manbarklak | snakewood |

**Resistant**

| | |
|---|---|
| acapu | kautaballi |
| 'cedar' | marish |
| kabukalli | pakuri |
| kauta | wallaba |

249

## Moderately resistant

baguacu
basralocus
mahogany, Brazilian

mandio
quaruba
suradan

## Very susceptible to attack

baromalli
dukali
heavy virola
kurokai
light virola

manniballi
Santa Maria
simaruba
sterculia

*The above classification is based on results of tests carried out in Puerto Rico and Trinidad, and refers to resistance to attack by both subterranean and dry-wood termites. Where the resistance to either type of pest differs, the lower rating is given.

# TURNERY

andiroba
boxwood
canafistula
gonçalo alves
grumixava
kingwood
louro, red

pakuri
pau marfim
purpleheart
satiné
sucupira
tulipwood

# VEHICLE BODIES

coigue
freijo
rauli

ulmo
vinhatico

# VENEER AND PLYWOOD

**Corestock**

andiroba
'cedar'

Chile pine
hevea
hura
'Parana pine'

## Decorative

arariba (selected)
andiroba
'cedar'
gmelina (selected)
imbuia
louro, red

mahogany, Brazilian
peroba rosa
purpleheart
rosewood
Santa Maria
satiné

## Utility (Plywood, chip-baskets, small laminated items, etc.)

Utility
andiroba
baromalli
courbaril (sapwood)
gmelina
hevea
hura

light virola
mahogany, Brazilian
manniballi
pakuri
'Parana pine'
sande

## AMENABILITY OF HEARTWOOD TO PRESERVATIVE TREATMENT

**Extremely resistant**

angelim pedra
'cedar'
greenheart
ipé
kurokai

louro, red
massaranduba
mora
purpleheart
Santa Maria
wallaba

**Resistant**

balsa
coigue
coubaril

mandio
quaruba
sterculia
white peroba

**Moderately resistant**

dukali

heavy virola
manniballi
rauli

**Permeable**

baromalli

ceiba
simaruba
light virola

# AMENABILITY OF HEARTWOOD TO PRESERVATIVE TREATMENT

The above classification refers to the ease with which a timber absorbs preservative under both open-tank (non-pressure) and pressure treatments. Sapwood, although nearly always perishable, is usually more permeable than heartwood, accordingly, the above classification refers to the relative resistance of heartwood to penetration.

### Extremely resistant
Timbers that absorb only a small amount of preservative even under long pressure treatments. They cannot be penetrated to an appreciable depth laterally, and only to a very small extent longitudinally.

### Resistant
Timbers difficult to impregnate under pressure and require a long period of treatment. It is often difficult to penetrate them laterally more than about 3mm to 6mm.
Incising is often used to obtain better treatment.

### Moderately resistant
Timbers that are fairly easy to treat, and it is usually possible to obtain a lateral penetration of the order of 6mm to 18mm in about 2-3 hours under pressure, or a penetration of a large proportion of the vessels.

### Permeable
Timbers that can be penetrated completely under pressure without difficulty, and can usually be heavily impregnated by the open-tank process.

# 3
# SOUTHERN ASIA

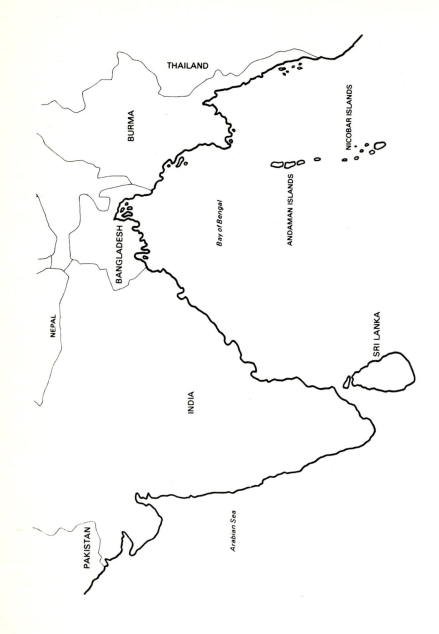

THAILAND

BURMA

NICOBAR ISLANDS

BANGLADESH

ANDAMAN ISLANDS

Bay of Bengal

NEPAL

SRI LANKA

INDIA

Arabian Sea

PAKISTAN

# INTRODUCTION

*The geographical scope of this chapter is confined to that area of Asia bounded on the north-north-east by the Himalayas, on the north-north-west by Afghanistan and Pakistan, and extending eastwards to the Yunnan and Shan mountain labyrinth, and southwards to the Mergui Archipelago. It includes the great land mass of India, Burma, Tenasserim, and Thailand, the Andaman Islands and Sri Lanka.*

*The entire area is situated near the centre of the southern part of the continent of Asia, but since from the southern extremity of India and Sri Lanka there is no land barrier between it and the South Pole, it is considered appropriate to regard the area as southern Asia since by this means the contents of this chapter can more readily be linked with those of the following chapter on timbers of South East Asia since there is, in many cases, a close botanical association between the timber species indigenous to both areas, which is not necessarily the case elsewhere in Asia.*

*The area covered by this chapter lies partly within, and partly without, the tropics, and contains many climates, often with wide variations of seasonal precipitation and temperature, all of which play a part in the successful, or otherwise, felling, drying, storage and marketing procedures of timber produced in the various countries.*

## India

*The forests of the western Himalayas, situated in the United Provinces, are mainly composed of coniferous species, although a variety of broad-leaved trees such as poplar, elm, birch, willow, etc occur, but the most important conifer in this area is the cedar or deodar, associated with blue pine, spruce, cypress, and occasionally silver fir.*
*At the lower limits of these forests, long-leaf pine, Pinus longifolia, sometimes forms pure stands, progressing gradually to the foothill and plains forests.*

*In the eastern Himalayas, deodar does not occur and gives way to spruce and silver fir at relatively low elevations, generally mixed with yew, juniper and larch, while lower still, the most prevalent species are oak and chestnut.*

255

The sal forests are found in two belts, one extending along the foot-hills of the Himalayas, the other occurring southwards through the Central Provinces to Madras.

On the hills and slopes of the Western Ghats are found the best teak forests, in contrast to the very dry, somewhat open teak forests of the Central Provinces.

Rainfall varies enormously in different localities; generally, in most localities except in practically rain-less tracts, there is a wet period from May or June to October, where rain is general according to the activity of the monsoon, and from November onwards for five months or so, very little rain is expected. Temperatures vary according to latitude; in November, north of 25° N it averages 18° C and in May from 27° C to 32° C, while below this latitude it is warmer, and towards the centre of the country it rises to a mean average annually of around 35° C with parts of Sind recording shade temperatures of 54° C.

The areas of greatest rainfall are on the crest of the Western Ghats from Bombay to the south of Calicut, in the hills of Assam, and from Calcutta along the coastal regions down the Bay of Bengal, with an annual precipitation of 5000mm and more in some localities.

## Burma

The most valuable and largest forest areas occur in Burma, and are fairly evenly distributed over the whole county, but teak takes pride of place, the forests producing this timber being found in Pegu Yomas, Martaban, the eastern slopes of the Arakan Yomas, the Chindwin Valley, and elsewhere in Upper Burma and the Shan States.

Climatic conditions vary very considerably in Burma; there are four rainfall zones, Upper Burma dry zone, 800mm at Mandalay; Upper Burma wet zone, 1800mm at Bhamo; Lower Burma littoral, 4800mm at Moulmein, 4200mm at Mergui, and 5800mm at Tavoy; sub-deltaic, 2400mm at Rangoon, and 960mm at Thayetmyo. The wet season is May to October, the dry season November to April, except on the coast and in the Kachin hills. Typical January and May temperatures are: Bhamo 17° C, 28° C., Mandalay, 22° C, 32° C; Mergui 26° C, 28° C; Rangoon 25° C, 30° C.

## The Andamans

The forest types found in the Andamans often contain species not found elsewhere, particularly the mangrove forests which fringe the shore within reach of the tide. The littoral forests, above tidal level, give way to evergreen forests generally found in valleys on low, alluvial soil, and moist, deciduous forests containing padauk, albizia, canarium, bombax, and other types.

## Sri Lanka

Sri Lanka is an island which slopes upwards on all sides from the coast towards the central massif. On the lowlands the temperature varies little throughout the year, and is less oppresive than on the Indian plains; rainfall is 3000mm gradually declining on the slopes. The forests yield satinwood and ebony, but generally the island is not considered a timber exporting source outside its immediate environs.

## Thailand

Thailand lies between Burma on the west and Laos on the east; the northern and western parts of the country are mountainous and covered by forests which yield a variety of hardwoods, the most important of which is teak. Much of Thailand's trade is with Singapore, Hong Kong and India, but timber and other materials are also shipped to Europe and America.

# PART I HARDWOODS

## AINI

*Artocarpus hirsuta* Lamk.          Family : Moraceae

**Other names**
anjili, ainee, pejata.

**Distribution**
Found on the west coast of India in evergreen forests.

**The tree**
A large to very large tree, attaining a height of 30m and a
diameter of 1.2m or more, with a straight, cylindrical bole, clear
of branches for about 21m to 24m.

**The timber**
The sapwood is white in colour, from 25mm to 50mm wide, and
the heartwood is golden-yellow when freshly cut, turning to
old gold or russet colour on exposure. The wood often has
darker lines, which after exposure attain a dark walnut-brown,
or blackish-brown colour. The grain is straight to interlocked,
and the texture is coarse, but even. The wood is light in weight,
weighing about 608 kg/m$^3$ when dried.

**Drying**
Dries easily and well without serious degrade developing.
Shrinkage, green to 12 per cent moisture content, 5.3 per cent
tangentially, 3.4 per cent radially.

**Strength**
Compared with teak, aini is about 7 per cent weaker in transverse strength, a little below teak in modulus of elasticity, but
about equal in compression parallel to the grain, and to shear in
a tangential direction, but is about 50 per cent more resistant
to shear in a radial direction.

**Durability**
Very durable.

## Working qualities
Easy to work with both hand and machine tools, and finishes with a lustrous surface.

## Uses
Furniture, cabinet-making, joinery, boat and ship building, turnery, cooperage, and decorative veneer from selected stock. Aini is regarded in India as an alternative to teak.

# AMARI

*Amoora wallichii* King,                    Family: Meliaceae
syn. *Amoora spectabilis* Miq.

## Other names
lalchini, galing libor.

## Distribution
Amari occurs in the Andamans, Burma, extending into Bangladesh, and in north-east India. It is found generally in moist, evergreen hill forests.

## The tree
A large tree, attaining a height of 18m to 27m with a straight and fairly cylindrical bole, usually about 6m in length, but occasionally up to 12m. The diameter varies but on average is about 0.5m.

## The timber
The sapwood is light-red in colour, while the rather dull heartwood is red when freshly cut, turning reddish-brown after drying. The wood is fairly soft, very straight-grained, and with a medium texture, and weighs about 528 kg/m³ when dried.

## Drying
Dries well, with drying characteristics similar to those of African mahogany (*Khaya* spp.).

## Strength
No information.

## Durability
Moderately durable.

## Working qualities
Easy to saw and work, and finishes to a fine, smooth surface. It takes the normal finishing treatments very well, and glues satisfactorily.

## Uses
Boat building, canoes and oars, furniture, cabinet-making, table and counter tops, and high-class joinery. It produces good veneer for plywood, and when selected, can be mildly decorative, especially when sliced from quarter-sawn flitches.

# AMOORA

*Amoora rohituka* Wight & Arn          Family : Meliaceae

## Other names
lota amara, galingasing (India) ; thitni, chaya-kaya (Burma).

## Distribution
Found in the western and north-eastern provinces of India, and from Nepal, through Bangladesh into Burma and the Andamans. It also occurs in Tenasserim, together with a similar and associated species, *A. polystachya*, which produces commercial tasua from Thailand.

## The tree
The tree varies with the locality, but on average produces a fairly straight, clean, cylindrical bole, some 6m or more in length, with a diameter of about 0.5m.

## The timber
The sapwood is light-red in colour, and the heartwood is red or wine-red when freshly cut, turning a deep reddish-brown on exposure. The wood is more coarse textured than amari (*A. wallichii*) due to the relatively larger pores, and the grain varies from straight to interlocked. It weighs about 560 kg/m$^3$ when dried.

260

## Strength
No information.

## Drying
Reputed to dry easily and well, but in some parts of India the tree is either ring-girdled, or is stored in water prior to conversion, which suggests that unless protected from severe drying conditions, the wood is liable to degrade.

## Durability
Moderately durable.

## Working qualities
The timber is said to saw and work well both by hand and on machines. It turns easily and smoothly, takes a good finish and a high polish.

## Uses
Canoes, knees for boats, joinery, and possibly for plywood.

# ANAN

*Fagraea fragrans* Roxb.                    Family : Loganiaceae

## Other names
Burma yellowheart.

## Distribution
Found in plains and forests south-east of Rangoon, in Martaban and Tenasserim, often in marshy localities.

## The tree
A moderate-sized tree with a stem some 9m to 18m high, and a diameter of 0.6m.

## The timber
The sapwood is pale yellow, and the heartwood yellowish-brown to light brown, turning the colour of old gold after exposure. Plain-sawn surfaces display a distinct figure caused by lighter coloured bands of parenchyma tissue alternating with broader, darker zones of fibrous tissue. The grain varies

from straight to irregular and deeply interlocked, the texture is medium-coarse but even, and the wood has a smooth oily feel. It weighs about 816 kg/m³ when dried.

## Drying
No information.

## Strength
Little is known of its mechanical properties. It is reputed to be about 10 per cent stronger than teak in compression parallel to the grain, but some 30 per cent weaker than teak in transverse strength.

## Durability
Very durable.

## Working qualities
It saws without difficulty, but requires care to obtain a really good finish.

## Uses
General construction, piling, bridges, boats and railway waggon bottoms.

# ANJAN

*Hardwickia binata* Roxb.                    Family: Leguminosae

## Other names
kamra, chhota dundhera, yepi, acha.

## Distribution
Widely distributed in the Central Provinces of India, often in fairly extensive gregarious patches.

## The tree
A large tree with a cylindrical and slightly tapering bole of about 9m and commonly 0.6m in diameter.

## The timber
The white, narrow sapwood is clearly demarcated from the

heartwood which is dark red to dark brown with darker streaks and often with a purplish cast. The pores and other elements contain copious amounts of dark reddish-brown gum. The wood is coarse textured, the grain is irregularly interlocked, and it weighs about 1075 kg/m³ when dried. The gum deposits impart an oily feel to the wood.

### Drying
A refractory species, difficult to dry, and liable to develop cracks in echelon following the interlocked grain. Cupping of relatively thin boards is also a problem.

### Strength
No data are available, but the wood is hard, heavy and strong.

### Durability
Durable.

### Working qualities
An extremely difficult timber to saw and work when dried, due to its hardness and gum content.

### Uses
Heavy construction, agricultural implements, oil mills, bridges.

## AXLEWOOD

*Anogeissus latifolia*, Wall.          Family : Combretaceae

### Other names
dhaura, dindal, dhawa, yerma, chiriman, golia.

### Distribution
Widely distributed in India, but is not found in the Andamans or in Burma.

### The tree
Although the various species of *Anogeissus* produce large, or moderate-sized trees under favourable conditions, they may be little more than shrubs in hot, dry locations. The best conditions for the growth of *A. latifolia* are in the sub-montane forests of

263

the United Provinces, and in the north and east of Kanara in Bombay, where the trees can produce clear boles of 9m to 12m in length, with a diameter of 0.6m or more.

## The timber
The wood is yellowish-grey, or uniformly grey, but as with other genera and species of the Combretaceae, there is usually a greenish or olive cast in the colour of the wood. There is no clear distinction between sapwood (which is wide), and heartwood, but in some logs, a small, irregular-shaped, purplish-brown heartwood is present.
The grain is shallowly interlocked in narrow bands, and sometimes is curly-grained in the radial plane, presenting an attractive fiddle-back mottling. The texture is fine, the wood is hard and heavy, and weighs on average 960 kg/m³ when dried.

## Drying
Difficult to dry from the green, there being a marked tendency for numerous fine wavy checks to develop on the surface, and for end splitting to occur. This to some extent can be attributed to dry, hot, air conditions prevailing in the areas of growth. It has been found that ring-girdled trees, left for eighteen months prior to felling, did not display such a high tendency to splitting and checking, and furthermore, green stock, kiln dried at Dehra Dun, was likewise relatively low in degrade. With suitable air drying techniques, and girdled trees, the degrade arising from rapid drying should be overcome.

## Strength
Axlewood is a strong, elastic, very tough and hard timber, comparing favourably with European ash in all strength categories, but being rather heavier than ash.

## Durability
Moderately durable.

## Working qualities
Fairly hard to saw, but it machines quite well and is capable of a smooth finish. It is a good turnery wood, finishing with a smooth shiny surface.

## Uses
Highly esteemed in India for agricultural implements and tool handles. It is also used for picking arms, cart wheels, spokes and axles, and for furniture.

Note. A closely allied species, *Anogeissus acuminata*, Wall, occurs in the teak forests of Burma, in Thailand, and sparsely in India. This produces commercial yon, a timber with a wide, whitish-grey to pale greenish-grey sapwood, often with yellow streaks or lighter grey bands, turning light greyish-brown on exposure, and a small, irregular, chocolate-brown heartwood. The wood, which may be a little lighter in weight than axlewood, and a little less hard, is non-durable, but accepts preservatives readily, and while not as popular in Burma as axlewood is in India, is nevertheless considered by Dehra Dun to be superior to axlewood, particularly as an alternative to American hickory for sucker rods in oil wells.

## BABUL

*Acacia arabica*, Willd.                    Family : Leguminosae

### Other names
jali, babbar, bamura, tuma, babli, kikar.

### Distribution
Indigenous to the Central Provinces and North Deccan in India, it is also largely planted in the drier parts of India, and in Burma, mainly as a road-side shade tree. Sometimes the tree forms pure forests in India, particularly in Bombay and Madras, while good plantations also exist in East Khandesh and Poona.

### The tree
A moderate to large-sized tree, attaining a height of 15m to 18m with a clear bole of about 7m and a diameter of 0.75m on the best sites such as Sind, but in the Deccan and Madras, it rarely attains a height of over 9m.

### The timber
The sapwood is wide, and whitish in colour, the heartwood is pinkish-white to light red when freshly cut, turning reddish-brown after exposure, usually mottled with darker streaks. The

265

wood is dull, with a rough feel, due to the coarse texture, and the grain is fairly straight, but may be twisted. It weighs about 816 kg/m³ when dried.

## Drying
Despite the dry, hot localities in which it grows, babul dries satisfactorily without undue degrade.

## Strength
Babul compares closely in general strength properties with keruing, being roughly equal in modulus of rupture and compression strength, but is a little lower in modulus of elasticity, and slightly superior in shear and hardness on side grain.

## Durability
Moderately durable to very durable depending on proportion of sapwood present.

## Working qualities
Works and machines reasonably well, and is capable of a good smooth finish, but the gummy cell contents give some trouble in sawing. It takes polish reasonably well, but due to the large pores requires a lot of filling.

## Uses
Construction, agricultural implements, dies, pit props, sleepers, anvil blocks, knees for boat building, oars, sugar and oil presses, carving, hookah stems and turnery.

# BENTEAK

*Lagerstroemia lanceolata* Wall.          Family : Lythraceae

## Other names
nana, vevala.

## Distribution
Found mainly in the Kanara and Malabar mixed deciduous forests of India.

## The tree
Moderate to large-sized tree, with a clean, cylindrical bole about 10m in length and 1.0m in diameter.

## The timber
The sapwood is greyish-white, and the heartwood is light red to reddish-brown, darkening to a uniform light walnut-brown. the wood is rather lustrous, with a fairly straight grain and a coarse texture, and weighs about 720 kg/m$^3$ when dried.

## Drying
Rather difficult to air dry, tending to warp and end-split if sawn green because of sun and hot winds. Girdling the trees a year before felling and prompt open piling of the stock under cover gives better air drying results. The timber kiln dries without undue degrade.

## Strength
A strong, elastic timber, equal to teak in crushing strength, superior in hardness, but slightly inferior to teak in compression parallel to the grain and in elasticity.

## Durability
Moderately durable.

## Working qualities
Fairly easy to work and machine, it finishes to a smooth surface, and takes a fair polish.

## Uses
Furniture, picking arms for textile mills, joinery, boat-building for hulls, masts and spars, railway waggon and truck bottoms, and building construction.


# BOMBAX, INDIAN

*Bombax malabaricum*, DC.                    Family : Malvaceae

## Other names
semul, cottonwood, letpan, simbal.

267

## Distribution
Very common and widely distributed throughout Pakistan, India, Bangladesh and Burma.

## The tree
A very large, straight, cylindrical-stemmed tree over 30m in height with a diameter of 0.75m or more above the buttress.

## The timber
The wood is pale yellowish-brown, often streaked with sap-stain; there is no distinction between sapwood and heartwood. The grain is straight, the texture very coarse but even, and the wood is soft and light in weight, about 400 kg/m$^3$ when dried.

## Drying
The timber dries rapidly and well, but is very prone to fungal attack. It should be converted green, immediately after felling, and then promptly stacked, in order to avoid fungal staining.

## Strength
A soft, weak wood, used for purposes for which strength is relatively unimportant.

## Durability
Perishable.

## Working qualities
Easy to saw and machine. There is difficulty in obtaining a smooth surface in planing, and a cutting angle of 20° gives better results.

## Uses
Boxes and packing cases, coffins, and plywood from rotary cut veneer.

An associated species, *Bombax insigne*, Wall. occurs in parts of India, Burma and the Andamans, and is known as didu. It is similar in appearance to Indian bombax, and is used for similar purposes, and in addition, for oar-blades in Burma.

# BOMBWAY, WHITE

*Terminalia procera* Roxb.                Family : Combretaceae

## Other names
badam, safed bombway.

## Distribution
Andaman Islands only.

## The tree
A tall tree with a long, straight bole, 12m or more in length, with a diameter of 0.6m to 1.0m.

## The timber
The sapwood is greyish in colour, often with yellow blotches which contain a dye that is soluble in water, while the heart-wood is light greyish-brown, somewhat lustrous, and very coarse-textured (the coarsest of all the Indian *Terminalias*). The grain is fairly straight, and the wood weighs about 660 kg/m$^3$ on average when dried.

## Drying
Dries fairly well, but with a slight tendency to distort and to surface check.

## Strength
Strength properties are generally similar to those of idigbo (*Terminalia ivorensis*) but bombway is a tougher timber, and rather harder.

## Durability
Non-durable.

## Working qualities
Works fairly easily, with only a moderate dulling effect on tools. A cutting angle of 15° reduces the tendency for grain to pick up during planing and moulding. The timber stains and polishes reasonably well, and takes nails and screws satisfactorily.

## Uses
Internal joinery, packing cases, and furniture from selected stock which can be quite attractive.

# BOW WOOD, ANDAMAN

*Sageraea elliptica* Hook f & Thoms.         Family : Anonaceae

## Other names
chai

## Distribution
An evergreen tree of Tenasserim and the Andamans.

## The tree
A small tree, producing cylindrical or very slightly fluted logs from 4.5m to 7.5m in length and 0.3m in diameter.

## The timber
Uniformly yellowish-white in colour, lustrous, with a smooth feel, fine and even-textured, straight grain, and weighing about 860 kg/m$^3$ when dried.

## Drying
Dries fairly well, but needs careful handling to avoid discoloration.

## Strength
No information, but the timber is reputed to be strong and elastic.

## Durability
No information.

## Working qualities
Works fairly easily, but there is a tendency for the grain to tear in planing.

## Uses
Used locally for boat ribs, and for native bows after steaming. It has been recommended for picking arms and striking tool handles.

# BOXWOOD

*Buxus wallichiana,* Baill.                    Family: Euphorbiaceae

**Other names**
chickri, shamshad, papri, sansadu.

**Distribution**
Found in the Himalaya region, and eastwards to Nepal.

**The tree**
A small tree which varies in size according to locality. In the Punjab it reaches 9m in height and 0.3m or a little more in diameter, with a clean stem of 3m to 4.5m.

**The timber**
There is little difference by colour between the sapwood and heartwood, the wood being a relatively uniform whitish-yellow to yellow, turning brownish-yellow after exposure. It is usually straight grained, but some logs may contain very irregular grain, and tension wood may be present. It has a silky lustre and an extremely fine and very even texture. The wood weighs about 830 kg/m³ when dried, rather less than European boxwood (*Buxus sempervirens*) which weighs 930 kg/m³.

**Drying**
Difficult to dry due to a pronounced tendency to split and surface check. One Indian method of reducing splitting is to make a single saw-cut down one side of the billet, as deep as the pith, in order to concentrate the shrinkage contraction during drying in the area of the cut. Boxwood dries slowly, and must therefore be protected from rapid drying conditions, preferably in cool, humid, and shaded, storage areas. Sawing the billet down the centre, and through the pith is advantageous, as is storage in water, or soaking the billet in a saturated solution of common salt, but this latter method should only be adopted where it is recognised that saws and equipment will be contaminated, and must be cleaned effectively after conversion of the wood.

**Strength**
No information.

### Durability
Durable.

### Working qualities
Although fairly resistant to sawing, it converts cleanly, and works to a smooth finish, although a reduction of cutting angle will reduce the tendency for grain tearing when irregular grain is present. The wood turns easily and well, and it can be stained and polished.

### Uses
Boxwood is used locally for the same purposes as European boxwood, for mallet heads, engraving, turnery, rules, protractors, musical instruments, fine cabinet work, carving and inlay. It is used in the hills for butter boxes, combs and snuff boxes.

# BULLET WOOD

*Mimusops elengi*, Roxb.　　　　　　　Family: Sapotaceae
and *M. littoralis*, Kurz.

### Other names
*M. elengi*; khaya mulsari, khaya (Andamans); kaya (Burma); elengi, malsari, bukal, borsali (India).
*M. littoralis*; mohwa (Andamans); katpali (Burma).

### Distribution
*M. elengi* is found in southern India, the Andamans, and in Burma in the Shan hills, Martaban and Tenasserim.
*M. littoralis* occurs in the Andamans, and on the Tenasserim coast in Burma.

### The tree
Both species attain heights of 21m to 36m with straight, cylindrical boles some 2.5m to 4.5m in length, and diameters of 1.5m or more.

### The timber
Both species are similar in appearance; the sapwood is pale

reddish to brownish-white, sharply defined from the heartwood which is rich deep red to dark reddish-brown, often striated with darker lines. *M. littoralis* is generally a little lighter in colour, and generally with straighter grain than that of *M. elengi*, in which the grain is often shallowly interlocked in narrow bands or is more or less irregular-grained. Both have a fine, even texture, and both weigh about 1075 kg/m³ when dried.

## Drying
Dries at a moderate rate with a slight tendency to warp and surface check in *M. elengi*, but with a pronounced tendency for distortion, and especially splitting and checking to develop in *M. littoralis*. Ring-girdling the trees is said to reduce this tendency in the latter species.

## Strength
*M. elengi* is stated to be a very hard, tough, and strong wood, some 50 per cent stronger than teak in compression parallel to the grain, while *M. littoralis* is about 16 per cent weaker.

## Durability
Durable.

## Working qualities
Rather difficult to work because of the general hardness of the wood, but a good surface is obtainable in planing, and *M. elengi* in particular takes a beautiful polish.

## Uses
Heavy construction, bridges, piling, turnery, and for boat-building for trenails, marlin spikes, belaying pins, spokes and handles of ships' wheels. Selected stock is used in India for furniture.

Note. An allied species, *Mimusops hexandra*, Roxb. occurs in central and southern India, and in Sri Lanka. The wood is similar to *M. elengi* in appearance, weighs about 1120 kg/m³ when dried, is difficult to dry unless ring-girdled, and is used locally for piling, agricultural implements, and general construction. It is known as palu, pala, manchi pala and rian.

# BURMA BLACKWOOD

*Dalbergia cultrata*, Grah.                    Family : Leguminosae

**Other names**
Indian cocobolo, yindaik, zaunyi.

**Distribution**
Found scattered and fairly commonly in dry deciduous forests throughout Burma.

**The tree**
A moderate-sized to large tree, with a straight, fluted stem about 7.5m long, and a diameter of about 0.6m.

**The timber**
The sapwood is greyish-white to pale brown in colour, and the heartwood brown with purplish streaks to purple or black streaks, with long, fine, parallel yellowish-brown lines. It is a very attractive wood, but less ornamental than Indian rosewood (*D. latifolia*). Ripple marks are present, and the wood has a fragrant scent, which however disappears after exposure. It is very heavy, weighing about 993 kg/m³ when dried, fairly straight grained, and medium to coarse textured.

**Drying**
The timber needs protection from rapid drying in order to avoid excessive splitting during air drying, but it is said to kiln dry reasonably well, but with a tendency for existing splits to extend.

**Strength**
No information.

**Durability**
Very durable.

**Working qualities**
Hard to work and machine, with considerable dulling effect on cutting edges. It turns well, and is capable of a good finish in planing, particularly if a 25° cutting angle is used. It takes a good and lasting polish.

## Uses
Decorative items and furniture, walking sticks, inlay and carving. It was successfully used as a substitute for lignum vitae in the lining of a stern tube of a tug in Rangoon. It had to be renewed after two years and three months, having worn down 12mm, whereas lignum vitae in the same tug had to be renewed annually, having worn down 16mm.

# CANARIUM, INDIAN

*Canarium euphyllum*, Kurz.                    Family: Burseraceae

## Other names
dhup, white dhup.

## Distribution
Andaman Islands.

## The tree
A very large tree with small buttresses, 24m to 30m high, yielding straight logs up to 12m in length and 1.0m or more in diameter.

## The timber
There is no distinction by colour between sapwood and heartwood, the wood being whitish to reddish or pinkish-grey, sometimes with a yellowish cast. It is lustrous, with a smooth feel, resembling gaboon (okoumé) in grain and texture, but appreciably lighter in colour. The grain is narrowly interlocked, and the texture is rather coarse. The wood weighs about 400 kg/m$^3$ when dried.

## Drying
A non-refractory timber, but liable to fungal staining if not converted green. It should be allowed to air dry rapidly. It is said to kiln dry without difficulty.

## Strength
In most strength categories it is about equal to American mahogany, but in bending strength along the grain, and in resistance to indentation, it is about 25 per cent inferior to mahogany.

## Durability
Perishable.

## Working qualities
Very easy to work by both hand and machine tools, but the interlocked grain is liable to pick up when planing or moulding quarter-sawn stock. With care, a smooth, shiny surface is obtainable in these operations. The wood takes nails and screws well, glues satisfactorily, can be stained and polished quite well.

## Uses
Interior joinery, light domestic flooring, packing cases.
An associated species, *Canarium strictum*, Roxb., occurs in the forests of the west coast of India. This is known locally as black dhup, or Indian white mahogany. The timber is light greyish white, with a pinkish cast, and weighs about 608 kg/m$^3$ when dried, being heavier and harder than white dhup. The timber is clean and straight grained, and of good appearance, but is difficult to dry, and its uses so far have been for packing cases and cheap planking.


# 'CEDAR, BURMESE'

*Toona ciliata* syn *Cedrela toona* Roxb.      Family: Meliaceae

## Other names
toon (India and Pakistan); yomhom (Burma); Burma cedar (UK).

## Distribution
India, Pakistan, Burma, Thailand, and in south-east Asia.

## The tree
In open situations the tree does not form a long bole; usually about 4m in length, but if drawn by other trees it forms a straight, cylindrical stem 9m to 12m to the first branch. The diameter can vary from 0.5m to 0.75m.

## The timber
The sapwood is pinkish or greyish-white, and the heartwood is

light brick-red when first exposed, turning a rich reddish-brown with darker streaks. The wood has a spicy odour, is somewhat lustrous, straight-grained, with a moderately close but rather uneven texture. It weighs about 450 kg/m³ when dried.

## Drying
The timber tends to warp, and thin sizes to cup, unless care is taken.

## Strength
A relatively weak wood.

## Durability
Moderately durable.

## Working qualities
The timber saws and works easily, but is inclined to be woolly. Occasional logs exhibit wavy grain and these contain irregular fibre alignment which is difficult to finish smoothly, especially with quarter-sawn stock. Good quality material takes a fine finish and polish. Selected logs yield decorative veneer.

## Uses
Furniture, interior joinery including doors, linings, carvings, cigar boxes, boat-building for masts and oars, cabins, etc. It is reputed to have a large movement value when subjected to fluctuating atmospheres, and this is considered a drawback to its use in the countries of origin.

# 'CEDAR, WHITE'

*Dysoxylum malabaricum,* Bedd.                    Family : Meliaceae

## Other names
bili-devdari, bili-budlige, vella, agil.

## Distribution
The trees are found in western India, especially in the Coorg, Mysore areas.

## The tree
A large tree with a straight, cylindrical stem, generally producing a clear bole some 9m in length and 0.5m or more in diameter.

## The timber
The sapwood is pale yellowish-white in colour, and the heart-wood is a light brownish-grey with a yellowish cast, turning darker on exposure. The wood is usually faintly striped along the grain with zones of darker tissue caused by parenchyma terminating the growth ring. It is lustrous, has a sweet scent, and a rather fine, even texture. The grain is straight or narrowly interlocked, and the wood weighs about 750 kg/m$^3$ when dried.

## Drying
The wood requires care in order to avoid twisting, warping, and splitting, and fungal staining of the sapwood. Green conversion, and protection from rapid drying conditions is required to avoid distortion, and careful sticking or anti-stain treatment to reduce or eliminate the staining.

## Strength
The strength properties are similar to those of teak, but about 20 per cent lower in compression parallel to the grain. It is equal to teak in modulus of elasticity.

## Durability
Very durable.

## Working qualities
The wood works easily and well with both hand and machine tools, and is a good turnery timber finishing with a shiny, smooth surface off the lathe. It takes the normal finishing media quite well.

## Uses
A high-class timber for furniture, planking, panelling, doors, window frames, flooring boards, cooperage, in which use, it is considered the best Indian wood for oil casks.

# CHAMPAK

*Michelia champaca,* Linn.                     Family: Magnoliaceae

## Other names
saga, sanga, sagawa (Burma).

## Distribution
Found in southern Burma and the west coast forests of India; cultivated extensively and often run wild.

## The tree
A large tree with a long, straight, cylindrical bole up to 21m in length and a diameter of 0.5m occasionally much larger.

## The timber
The sapwood is white, and the heartwood light yellowish-brown to olive-brown, lustrous, soft, straight grained, with a medium texture, similar to American whitewood (*Liriodendron tulipifera*), but considered to be finer textured. The wood weighs between 490 kg/m$^3$ and 544 kg/m$^3$ when dried.

## Drying
Air dries reasonably well if trees are ring girdled and left for three years before felling and conversion. Liable to split severely if converted green, even with care under local conditions.

## Strength
Similar to deodar in hardness, shear strength, and in crushing strength across the grain, but in other respects it is weaker than deodar.

## Durability
Non-durable.

## Working qualities
Easy to saw and work, finishing smoothly and taking a good polish. It is unsuited to turnery.

## Uses
Joinery, boat building for bent wood ribs, drawer linings for furniture, and veneer for plywood.

# CHAPLASH

*Artocarpus chaplasha*, Roxb.                    Family : Moraceae

## Other names
taung-peinne (Andamans and Burma).

## Distribution
Occurs from Nepal eastwards through Bangladesh, Assam, Bengal, into lower Burma and the Andamans.

## The tree
A large, or very large tree, producing clean, straight, cylindrical stems up to 18m to 21m in length and 0.75m to 1.2m in diameter.

## The timber
The sapwood is white or pale yellowish-white in colour, and may be 50mm wide, and the heartwood is yellowish-brown to golden brown with lighter and darker streaks, frequently marked with white lines caused by chalky deposits in the pores. The photo-chemical reaction is very marked and the wood darkens to brown or dark brown after only a few hours exposure. The grain is straight or shallowly interlocked, and the texture is very coarse but even. The wood weighs about 480 kg/m$^3$ when dried.

## Drying
Rather difficult to dry, tending to warp, split, and to develop surface checking.

## Strength
Reputed to have only moderate strength. Tests at Dehra Dun suggest it to be 26 per cent below teak in transverse strength, and 28 per cent weaker in compression parallel to the grain.

## Durability
Moderately durable.

## Working qualities
It saws easily and well, but owing to interlocked grain it requires care to obtain a good surface in planing and moulding.

Reducing the cutting angle to 20° reduces the tendency for grain tearing to occur. It polishes well, but owing to the markedly large pores requires a lot of filling. It peels well, and produces veneer suitable for plywood, and when selected, for decorative use.

## Uses
Boat building, masts and oars, joinery, doors, window frames, light construction, plywood and cheap furniture.

# CHICKRASSY

*Chukrasia tabularis,* Adr. Juss.          Family : Meliaceae

## Other names
Chittagong wood, lal devdari, yinma.

## Distribution
India, Bangladesh, Burma and the Andamans.

## The tree
A tall, long stemmed tree, with a clean, straight, cylindrical bole about 9m in length and 0.6m or more in diameter.

## The timber
The sapwood is pale yellowish or brownish-white merging into the heartwood which is yellowish-red to red when freshly cut, turning yellowish or reddish-brown after exposure. The wood has a beautiful satin sheen, is medium fine-textured, and the grain is irregularly interlocked or wavy. It weighs about 640 kg/m$^3$ when dried.

## Drying
Dries rapidly and well, with only a slight tendency for fine surface checking to develop.

## Strength
Similar to European oak in all strength categories.

## Durability
Non-durable.

## Working qualities
A relatively easy wood to work by hand and with most machine tools; the irregular grain is liable to tear out during planing and moulding, but a reduction of cutting angle to 20° is helpful. It takes nails and screws well, and stains and polishes excellently. It peels well and is capable of producing good, decorative veneer.

## Uses
Furniture, carving, panelling and high-class interior joinery.

## CHUGLAM, BLACK

*Terminalia manii*, King.                    Family: Combretaceae

## Other names
kala chuglam.

## Distribution
Found in the Andamans.

## The tree
A tall tree, producing long, straight boles, from 9m to 12m long, and 0.75m in diameter.

## The timber
The sapwood is greyish-white with a faint yellowish cast, and the heartwood is olive-grey when freshly cut, turning to brownish-grey, sometimes with purplish-black streaks. The wood is straight grained, medium-fine textured, and weighs about 816 kg/m³ when dried.

## Drying
Fairly difficult to dry in hot, dry climatic conditions, since the wood tends to split and check under these conditions. Green conversion, and prompt open stacking under cover gives the best results.

## Strength
The general strength properties are similar to those of teak, but a little inferior to that timber in compression parallel to the grain.

## Durability
Probably durable.

## Working qualities
Moderately easy to work. It planes to a good smooth surface, but is a little inclined to pick up. It has good nailing and screwing properties and takes polish well with little filling.

## Uses
High-class joinery, staircases, laboratory tables, flooring and light construction.

# CHUGLAM, WHITE

*Terminalia bialata,* Steudel          Family : Combretaceae

## Other names
Indian silver grey wood, chuglam, lein.

## Distribution
Found in the Andaman Islands.

## The tree
A very large tree attaining a height of 30m to 48m and a diameter of 0.75m to 1.5m.

## The timber
In most logs there is no distinction by colour between sapwood and heartwood, the wood being a uniform greyish-yellow colour, but in selected logs there is a false heartwood of light, nut-brown or olive-brown colour banded with darker streaks, producing the highly ornamental silver grey wood. The wood is lustrous, has a smooth feel, and weighs about 690 kg/m$^3$ when dried. Both types of timber have fairly straight grain, and medium texture.

## Drying
There is a tendency for end splits and surface checks to develop during air drying under local conditions. With care, degrade is likely to be minimal. Green conversion, and prompt storage under cover is recommended.

## Strength
Similar to European oak (*Quercus* spp.) in bending strength and compressive strength along the grain, resistance to suddenly applied loads and resistance to indentation, but is a little stiffer than oak.

## Durability
Probably moderately durable.

## Working qualities
The wood works with moderate ease, and is capable of a smooth silky finish. It holds nails and screws well, and can be glued, stained and polished satisfactorily. Can be sliced for decorative veneer.

## Uses
The decorative silver grey wood can be used for all decorative work, panelling, furniture, cabinet making, and high-class joinery, and the plain chuglam for joinery, mathematical instruments, boat fittings, staircases.

# CINNAMON

*Cinnamomum tavoyanum,* Meissn.          Family: Lauraceae

## Other names
hmantheinpo.

## Distribution
Found in South Tenasserim.

## The tree
A large tree producing a clean, cylindrical bole some 12m in length and 1.0m in diameter.

## The timber
There is no clear distinction by colour between sapwood and heartwood, the wood is light brownish-grey to olive grey, often with a pinkish tinge toward the centre of the log. It has a strong aromatic scent which is retained for a long time, and the wood contains pith flecks caused by the mining of the larvae of

Diptera beetles which burrow in the cambium. The grain is straight, and the texture medium-fine and even. The wood weighs about 608 kg/m³ when dried.

## Drying
Reputed to dry well without undue degrade.

## Strength
No information.

## Durability
Durable.

## Working qualities
Easy to saw, and works well, both by hand and with machine tools, finishing to a smooth shiny surface.

## Uses
A close straight grained wood, free from knots, and of good appearance, suitable for joinery, shop-fitting, doors, light construction, boat building.

# CUTCH

*Acacia catechu,* Willd.                    Family : Leguminosae

## Other names
khoira, kagli, sha, karan-galli.

## Distribution
Found in India, West Pakistan, Bangladesh and Burma.

## The tree
A relatively small tree producing a moderately straight and cylindrical bole 2.5m to 3.0m or a little more in length, with a diameter of about 0.3m.

## The timber
The sapwood is wide, yellowish-white in colour, clearly demarcated from the heartwood, which is light or dark red when freshly cut, turning to brownish-red after exposure. The

pores contain abundant reddish-brown gummy infiltrations and chalky deposits. The wood is somewhat lustrous, with a smooth feel, the grain is straight, and the texture medium to coarse. It weighs about 1000 kg/m³ when dried.

## Drying
The timber dries slowly, but well, with a tendency for the ends to split in thick stock.

## Strength
An extremely strong and very hard timber, but there are no data available.

## Durability
Very durable.

## Working qualities
Reputed to be a difficult timber to saw and machine.

## Uses
Essentially the wood is used locally for chipping from which cutch (a brown dye) is extracted, but the timber is also used extensively for construction, for wheel felloes, tent pegs, oil and sugar cane crushers, mallet heads and plane bodies, tool handles.

# DHAMAN

*Grewia tilaefolia,* Vahl.                    Family : Tiliaceae

## Other names
dhamani, damnak, unu, pharsia.

## Distribution
Found mainly in northern and central India.

## The tree
A moderate to large-size tree, usually producing a bole some 4.5m to 6m long and 0.5m in diameter, but in some areas, especially in Malabar, boles of 9m and a diameter of 0.75m or more are common.

## The timber
The sapwood is pale yellowish-white, and the heartwood is reddish-brown to brown in colour, with darker streaks caused by seasonal zones of dark fibrous tissue, and often marked by white spots. The grain is straight or wavy, and the texture is medium. The wood weighs about 736 kg/m$^3$ when dried.

## Drying
Green conversion and careful treatment by sticking under cover gives good results. It is liable to end splitting and surface checking unless care is taken.

## Strength
The timber has about the same toughness as teak, but is 20 per cent stiffer, has 10 per cent greater resistance to crushing and is much harder than teak.

## Durability
Very durable.

## Working qualities
Is said to work easily both by hand and machine tools, taking a good finish in planing and moulding if cutting angles are reduced to 20°. It takes a high polish.

## Uses
Shuttles, bobbins, and picker arms in textile mills, and is considered superior to haldu (*Adina cordifolia*) for these purposes because of its better stability in changing climatic conditions. Also used for kegs and barrels, and for cart construction.

## EBONY

*Diospyros species*                    Family : Ebenaceae

The trade name ebony covers all species of the genus *Diospyros* with predominantly black heartwood, and accordingly, *D. ebenum* (Ceylon ebony) with its more consistent production of black wood has been termed true ebony. There are however, other species whose wood may be wholly black or only slightly streaked, or may be consistently streaked with black, brown, or grey. The following is a description of the various species.

***Diospyros ebenum***, Koenig, known as true or genuine ebony.

## Other names
tendu, tuki, ebans.

## Distribution
Sri Lanka and southern India.

## The tree
*D. ebenum* reaches its best development in Sri Lanka, where the diameter is usually about 0.75m whereas in India it is usually 0.3m or a little more. Good straight boles are generally about 4.5m in length.

## The timber
The sapwood is light yellowish-grey, and the heartwood is jet black, very rarely with a few lighter streaks. It has a metallic lustre and a fine even texture; the grain may be straight or somewhat irregular or wavy. Weight about 1190 kg/m³ when dried.

## Drying
Generally very difficult to dry as the wood tends to develop long, fine, deep cracks, especially if cut into large sections. The logs can be converted green, and then sawn to the smallest permissible sizes, placed in stick and protected from hot sun, or the trees can be girdled two years prior to felling, the converted stock being stored under cover.

## Durability
If the sapwood is removed, the black heartwood is very durable.

## Working qualities
Extremely hard to work both by hand or machine, the black material being rather brittle and inclined to chip. Irregular or wavy grain tends to pick up in planing, and cutting angles need to be reduced to 20° in order to obtain a smooth finish. It turns well and with care finishes to a shiny dense surface. It takes a high polish.

288

## Uses
Fancy goods, parts for musical instruments eg piano keys, finger boards, tail-pieces and saddles for stringed instruments, and for turnery, cutlery handles, brush-backs, chop-sticks and for inlay.

*Diospyros tomentosa* Roxb, known as ebony, and also as tendu, kendu and temru.

## Distribution
Found in the United Provinces of India and through to Nepal.

## The tree
In favourable locations grows to a moderate size with a bole of about 3.5m in length and a diameter of 0.3m but is little more than a shrub in hot, dry localities.

## The timber
The sapwood is wide, pinkish-grey to light or dark brown in colour, sharply defined from the heartwood, which rarely exceeds 150mm in diameter, roughly about half the diameter of the log, is irregular in shape, and black in colour, but is often streaked with brown or purple irregular bands. It is fine and even textured, with a straight or wavy grain, and weighs about 848 kg/m$^3$ when dried.

## Drying
Air dries well provided it is protected from hot sun. The heartwood tends to develop long splits.

## Uses
The heartwood is extremely hard, but brittle, while the sapwood is extremely elastic and strong. Very durable, the dark wood is used for small decorative articles, cabinet work, brush-backs, and carved walking sticks, while the light-coloured wood is used for tool handles, picking arms, cart shafts and spokes for wheels.

*Diospyros melanoxylon* Roxb. Also produces ebony, also known as tendu, tunki, timbruni and temru.

## Distribution
Found in the Central Provinces of India and southwards to Sri Lanka.

## The tree
In the scrub jungle the tree is little more than pole-size, but on favourable sites it has a straight stem 4.5m to 6.5m in length with a diameter of 0.75m or more.

## The timber
The sapwood is wide and pinkish-grey to pinkish-brown in colour, sharply defined from the small, irregularly-shaped heartwood which rarely exceeds 200mm in diameter, and which is black in colour, often streaked with lighter bands. It has a fairly straight to curly grain, is medium-fine and even textured and weighs about 880 kg/m$^3$ when dried. The dark coloured wood is said to be very durable, and the lighter coloured sapwood as being durable.

## Uses
The heartwood is used for decorative items, snuff boxes, cabinet and inlay work, toys and combs, while the sapwood, now regarded in India as being as valuable as the heartwood, is used for picking arms, tool handles, mallets, plane bodies, and for many other uses which require strength, elasticity and a smooth finish. The light wood is further used for billiard cue shafts with the black wood being used for the butts.

A further species of *Diospyros* furnishes Andaman marblewood; this is described elsewhere in this booklet.


# ELM, INDIAN

*Holoptelea integrifolia*, Planch.          Family: Ulmaceae
syn. *Ulmus integrifolia*, Roxb.

## Other names
wowoli, tapsi, myaukseik, thale.

## Distribution
India, Bangladesh, Burma.

## The tree
A large to very large tree, attaining its largest size in Burma, with a straight, cylindrical bole up to 15m in length with a diameter

of 1.0m or more. In India it is generally smaller, producing a clear bole of about 6m in length and a diameter of 0.6m to 0.75m.

## The timber
The wood is a uniform light yellow in colour, with white vessel lines along the grain caused by chalky inclusions in the pores. The wood may sometimes appear to have a greyish tinge, but when present is due to discoloration caused by fungal stain, the wood containing a very high starch content. It is rather lustrous, has a medium and even texture, and the grain is interlocked in broad, straight bands. It weighs about 624 kg/m$^3$ when dried.

## Drying
Air dries and kiln dries easily and well, but is very prone to fungal staining, and must be allowed to air dry as quickly as possible after conversion.

## Strength
Moderately soft but relatively strong, it is about 25 per cent weaker than teak in elasticity, and about 9 per cent below that species in transverse strength.

## Durability
Non-durable.

## Working qualities
Saws and machines reasonably well, but there is a tendency for the interlocked grain to present some difficulty in planing and moulding. It takes nails and screws well, and can be polished satisfactorily after suitable filling.

## Uses
Joinery, door frames, panelling, warper bobbins in jute mills, cotton reels, carpentry and cheap furniture.

## ENG

*Dipterocarpus tuberculatus,* Roxb.　　Family : Dipterocarpaceae

## Other names
in (Burma) ; pluang (Thailand).

## Distribution
Burma and Thailand.

## The tree
A large tree with a straight cylindrical stem, which on favourable sites reaches 15m to 24m to the first branch, but in less favourable locations the tree is much shorter.

## The timber
The sapwood is greyish or reddish-white, and the heartwood is reddish-brown, turning darker on exposure, with short, whitish, tangential lines at irregular but relatively close intervals, caused by longitudinal resin canals. The wood is dull, has a rather rough feel, due to the coarse texture. fairly straight to somewhat interlocked grain, and weighs about 880 kg/m³ when dried.

## Drying
Slow drying, with a tendency to warp, check and split and to shrink excessively. Storage of the logs for a year prior to conversion is said to reduce air drying degrade.

## Strength
Although heavier than gurjun (*Dipterocarpus* spp.) it is about equal to that timber in bending strength and stiffness, but is about 50 per cent harder, and 40 per cent more resistant to shock loads.

## Durability
Moderately durable.

## Working qualities
Fairly easy to work and finish. Takes a fairly smooth finish, screws and nails very well, and can be polished, but not as satisfactorily as gurjun.

## Uses
Heavy construction, sleepers, flooring, railway waggon floors.

# GRANDIS GUM

*Eucalyptus grandis* Hill ex Maio          Family : Myrtaceae

## Other names
grandis gum, blue gum rose gum, flooded gum.

*E. grandis* is very similar to *E. saligna* Sm. and the two species are considered together under the British Standard name of saligna gum.

## Distribution
Various species of *Eucalyptus*, although native to Australia, have been extensively planted elsewhere in the world, including southern Asia.

## The tree
A tall tree attaining a height of 43m to 55m and a diameter of 1.0m or more, with clean boles up to 24m in length. Heart rot tends to develop in old trees.

## The timber
The sapwood may be up to 50mm wide, and the heartwood is variable, ranging from pinkish-white to pink or rose-brown, depending on age and other characteristics. The grain is usually interlocked or may be straight, and the texture is medium to coarse. It weighs about 705 kg/m$^3$ when dry.

## Drying
Fairly difficult to dry, and requires care in the early stages to avoid surface checking. Cupping tends to occur in plain-sawn material, and timber from fast-grown trees tends to distort. Collapse, while liable, does not generally result in corrugated surfaces, and reconditioning is not usually required. Resawing after drying should be avoided, the best results being obtained from sizes as originally sawn, especially from quarter-sawn stock.

## Strength
Similar to jarrah.

## Durability
Non-durable.

## Working qualities
Works easily and well, and finishes satisfactorily. It holds nails well, but tends to split when nailing near the ends of pieces.

## Uses
Construction, boxes, weather boards and shingles when preservative treated, flooring.

# GUMHAR

*Gmelina arborea,* Linn.  Family: Verbenaceae

## Other names
gomari, shiwan, yemane, gambari, gmelina.

## Distribution
Widely distributed throughout India and Burma, but nowhere common in India, nor near the large rivers in Burma, due to heavy local demands, especially for boat building.

## The tree
Variable, according to locality. In Burma it attains a height of 30m with a clear bole of 9m and a diameter of 0.75m on average, but occasionally more. In India the diameter is commonly 0.45m.

## The timber
One of the best and reliable timbers found in southern Asia. The sapwood is not distinct by colour from the heartwood, the wood being yellowish-brown, lustrous, with a smooth, oily feel. The grain is somewhat irregular and interlocked, and the texture is medium-coarse. It is light in weight, about 480 kg/m$^3$ when dried.

## Drying
Dries well without undue degrade, and while the wood shrinks considerably during drying, it is remarkably stable when dried.

## Strength
Gumhar is botanically related to teak, and by comparison with this timber, it is not quite as strong, but has the advantage of

being some 160 to 190 kg/m³ lighter in weight. It is weaker than teak by about 17 per cent in transverse strength, 30 per cent in compression parallel to the grain, and 16 per cent in modulus of elasticity.

## Durability
Durable.

## Working qualities
Works readily to a fairly smooth finish, and takes stain and polish well.

## Uses
Boat decking, planking, musical instruments, carving, clogs and sandals, doors, panelling, furniture, light construction.

# GURJUN

*Dipterocarpus* spp.                    Family: Dipterocarpaceae

Various species of the genus *Dipterocarpus* occur widely throughout southern Asia; with the exception of *D. tuberculatus*, the eng of Burma, the weight of the principal timber produced by these species is similar, and about 740 kg/m³ when dried. With the further exception of *D. zeylanicus*, the hora of Sri Lanka, the species are grouped for commercial use into single descriptions, as follows,

*D. grandiflorus* Blanco, *D. indicus* Bedd, *D. macrocarpus* Vesque, are the principal species producing Indian or Andaman gurjun, with *D. pilosus* Roxb., known in India as hollong or gurjun.

*D. alatus* Roxb., *D. turbinatus* Gaertn f., are the principal species producing Burma gurjun, and also yang of Thailand, or Thailand gurjun.

*D. alatus* Roxb., *D. costatus*, *D. dyeri* Pierre, *D. intricatus* Dyer, and *D. obtusifolius* Teijsm ex Miq, are the principal species producing dau of South Vietnam and Cambodia.

295

### General characteristics
While variations in appearance and general characteristics exist according to locality, all have light red to reddish-brown or brown heartwood, and a lighter coloured sapwood which is usually 50mm to 75mm wide. The grain is straight or sometimes shallowly interlocked, and the texture coarse, but even.

### Drying
In general all these species dry slowly, with some reluctance to give up moisture uniformly, particularly in thick material. High temperatures in both air drying and in kilning tend to encourage the exudation of oleo-resin.

### Strength
In most strength properties the various species are comparable to teak, but are about 25 per cent stiffer, and about 40 per cent more resistant to shock loads.

### Durability
Moderately durable.

### Working qualities
The working properties vary, but in general there is a moderate to severe dulling effect on cutting edges, while in some instances resin can cause difficulties due to gumming up of tools and machine tables and fences. In planing and moulding, a good finish is obtained from straight grained material, but the interlocked grain is liable to pick up when working quarter-sawn stock, and the cutting angle should be reduced to 20° in order to improve the finish. The wood takes nails and screws fairly well, and can be stained without difficulty, but the resin content can prevent good results in polishing or varnishing.

### Uses
Heavy construction, sleepers, wharf decking, mill floors, vehicle and railway waggons. For the most exacting external uses the timber should be treated with preservatives.

## HALDU

*Adina cordifolia* Benth. & Hook, ex Brandis

Family: Rubiaceae

### Other names
hnaw (Burma) ; kwao (Thailand).

## Distribution
Found in India, Burma and Thailand.

## The tree
A very large tree with a long, straight, fluted stem and slightly buttressed base, often as much as 15m to the first branch. Logs of 1.5m diameter are not uncommon.

## The timber
The sapwood which is yellowish-white in colour and rather wide, merges gradually into the heartwood which is citron-yellow when freshly cut, turning pale yellowish or reddish-brown on exposure. A lustrous wood with a fairly smooth feel, it has a fine and even texture, and a grain that is fairly straight, but may be somewhat spiralled or interlocked in broad bands. It weighs about 670 kg/m$^3$ when dried.

## Drying
Air dries well, and kiln dries extremely well.

## Strength
A strong and hard wood, only slightly less strong than teak in compression, elasticity, and in transverse strength, but superior to teak in shear strength and hardness.

## Durability
Non-durable.

## Working qualities
It works fairly easily, and turns well, and takes stain and polish satisfactorily.

## Uses
Joinery, furniture, cabinet-making, jute bobbins, panelling, flooring (as a substitute for maple), carving.

# HOLLOCK

*Terminalia myriocarpa* Heurck and Muell                Family : Combretaceae

## Distribution
Found in Upper Burma and in the eastern Himalayas from Nepal to Assam.

## The tree
A very large tree with a 12m clean, straight bole, and a diameter of 1.0m but sometimes much more.

## The timber
The sapwood is light brown, and the heartwood is light brown when first cut turning darker with age. The wood is beautifully striated with narrow dark streaks; lustrous, with a coarse texture and a straight, but sometimes wavy grain, weighing 625 kg/m$^3$ when dried.

## Drying
Dries with little degrade.

## Strength
A strong timber equal to teak in hardness and to kindal (*T. paniculata*) in other strength properties.

## Durability
Non-durable.

## Working qualities
Saws and works reasonably well, but is rather hard to work with hand tools when dry. It machines without undue difficulty, but material with wavy grain tends to pick up during planing especially on quarter sawn surfaces. It peels readily on the veneer lathe.

## Uses
Plywood mainly, but suitable for interior joinery and construction.

# KINDAL

*Terminalia paniculata,* Roth.                  Family : Combretaceae

## Other names
pekarakai, marwa, neemeeri, honal.

## Distribution
Found mainly in the deciduous forests of the Western Ghats of India.

298

## The tree
A large tree producing 9m long clean, and fairly cylindrical boles with an average diameter of 0.5m or more in favourable localities.

## The timber
The sapwood is grey, and the heartwood greyish-brown to brown, turning darker with age. Fairly lustrous, with a straight grain and a medium texture. Moderately heavy, it weighs about 768 kg/m³ when dried.

## Drying
Under local conditions the timber tends to split excessively if left to dry in the log. Green conversion and immediate stacking in stick, with good top cover is reasonably satisfactory. Kiln drying from the green gives little trouble.

## Strength
A strong, and very hard timber, equal to teak in transverse strength and elasticity, but considerably harder and stronger in shear strength.

## Durability
Moderately durable.

## Working qualities
Moderately difficult to machine, with an appreciable dulling effect on cutters, and needing care in planing and moulding to obtain a good smooth surface. It has very good nail and screw holding properties, and can be stained and polished.

## Uses
It is used locally for construction, ladder strings, boats, carts, railway waggon flooring and doors, joinery and furniture.

## KOKKO

*Albizia lebbeck*, Benth.  Family: Leguminosae

## Other names
siris (India); 'East Indian walnut' (UK). The latter name is confusing and its use should be discontinued.

## Distribution
India, the Andaman Islands and Burma.

## The tree
A large tree with an umbrella-shaped crown, some 18m to 30m in height, and a diameter from 0.6m to 1.0m producing clear stems up to 9m in length.

## The timber
The wide sapwood is yellowish-white in colour, and the heartwood is dark brown streaked with lighter and darker bands. The grain is often deeply interlocked or wavy, giving the wood a figured appearance. The texture is medium to coarse, and the wood weighs about 650 kg/m³ when dried.

## Drying
Logs are liable to develop heart shakes unless converted green. Sawn stock must be protected from hot sun in order to prevent surface checking, and end sealing of thick material helps to reduce end splitting. Kiln drying of green stock can be carried out satisfactorily.

## Strength
The strength properties are similar to those of American black walnut (*Juglans nigra*).

## Durability
Moderately durable.

## Working qualities
Slightly hard to work by hand and machine tools. The interlocked grain tends to pick up in planing and moulding, particularly on quarter-sawn surfaces, and a cutting angle of not more than 20° should be used. It polishes excellently, but requires careful grain filling.

## Uses
Furniture, panelling, carving, boats, casks and for structural purposes.

# 'LAUREL, INDIAN'

*Terminalia alata* Roth.,                Family : Combretaceae
*T. coriacea* W & A
and *T. crenulata* Roth.

## Other names
sain, amari (India) ; taukkyan (Burma).

## Distribution
India, West Pakistan, Bangladesh and Burma.

## The tree
A very large tree on favourable sites sometimes attaining a diameter of 1.5m with a 21m clean, straight bole ; more commonly it is found with a 12m to 15m bole and a diameter of 0.9m.

## The timber
The sapwood is reddish-white, sharply defined from the heartwood which is variable in individual trees, ranging from light brown with few markings or finely streaked with darker lines, to dark brown or brownish-black and often banded with streaks of darker colour. The wood is fairly straight grained, coarse textured, and moderately heavy to heavy, weighing between 736 kg/m³ and 960 kg/m³ when dried.

## Drying
A highly refractory timber to dry being prone to surface-checking, warping and splitting. The timber should be converted green, preferably during the rainy season, and allowed to dry slowly in stick under top cover.

## Strength
Its strength properties are similar to those of English oak, but it is very much harder on side grain, and a little stiffer than oak.

## Durability
Durable.

## Working qualities
Fairly difficult to work with hand tools, and moderately hard to saw and machine; straight grained material finishes cleanly but when interlocked grain is present there is a tendency to pick up in planing and moulding, particularly on quarter-sawn surfaces, and a cutting angle of 20° is advisable. Hard to nail but holds screws well, and glues firmly. A fair amount of filling is needed for a high polish, and a wax finish gives better results.

## Uses
Furniture, cabinet making, brush-backs, panelling, boat building, high-class joinery, police batons, large piles in harbour work, tool handles and for veneer.

# LUNUMIDELLA

*Melia composita*, Willd.                    Family: Meliaceae

## Other names
Malabar nimwood, nimbarra, 'Ceylon mahogany', 'Ceylon cedar'. The latter two names are confusing and should be discontinued.

## Distribution
India, Bhutan and Sri Lanka.

## The tree
A large deciduous tree, up to 0.5m in diameter with a 9m clear, straight, cylindrical bole.

## The timber
The sapwood is greyish-white with a yellow cast, and the heartwood is light pink to light red when freshly cut, turning to pale russet-brown after exposure. The wood is lustrous, has a straight grain and a coarse and somewhat uneven texture, is soft and light in weight, about 340 kg/m$^3$ when dried.

## Drying
The wood is liable to fungal staining and end splitting. It should be converted green, and the timber piled in stick, preferably under cover.

**Strength**
A rather weak timber, but generally strong enough for the uses to which the timber is put.

**Durability**
Non-durable.

**Working qualities**
Very easy to work and machine.

**Uses**
Cigar boxes, musical instruments, out-riggers of native boats and for plywood.

An allied species, *Melia azedarach*, Linn. occurs in the forests along the base of the Himalayas, and all over India, often in irrigated plantations. It is known as Persian lilac, darachik, tamaga, etc, and is harder and darker in colour than *M. composita*. It weighs 576 kg/m$^3$ when dried, is durable, and is used for toys slips for tennis racquet handles, small boxes and for veneer. It is an attractive timber, but is restricted somewhat by the size of the logs which are about 3.5m long with a diameter of 0.3m or slightly more.


## MARBLEWOOD, ANDAMAN

*Diospyros marmorata* Park.                    Family : Ebenaceae

**Other names**
zebrawood, this name should be discontinued.

**Distribution**
Andaman Islands.

**The tree**
A moderate-sized tree, 12m to 21m in height, and 0.3m to 0.6m in diameter. The average size of bole is 2.5m to 3.0m in length and about 0.3m in diameter, with the darkly figured portion usually being about 150mm wide.

## The timber
The wood is generally light grey to greyish-brown, often with darker streaks or jet black bands; the darker bands sometimes alternate fairly uniformly with the lighter tissue, while in other cases they are very irregular and show on end grain surfaces as oval dark or black spots. The wood has a smooth feel, is fine and even textured, and the grain is fairly straight. It weighs about 1030 kg/m³ when dried.

## Drying
Difficult to dry, with a distinct tendency to warp and to develop end and surface checks. Requires to be converted from the green to the smallest permissible sizes and then allowed to dry slowly under top cover.

## Strength
A relatively strong timber, but its potential market depends on its decorative value, not its strength.

## Durability
Very durable.

## Working qualities
Difficult to saw and plane, but with care, is capable of a hard, smooth finish. It is considered first-class for turnery since it is not so brittle and liable to chipping out as true ebony. It takes a beautiful finish, but care is needed in the choice of media otherwise the colour contrast may be lost.

## Uses
Carving, turnery, cabinet work, inlay and small decorative items.

# MESUA

*Mesua ferrea,* Linn.                    Family: Guttiferae

## Other names
nahor, nangal, gangaw.

## Distribution
Found in tropical forests in Assam, south-west India, Upper Burma, Tenasserim, the Andamans, and in Malaya, where the timber is known as panagua.

## The tree
A fairly large tree with a straight stem, clear for upwards of 12m with a diameter of 0.5m.

## The timber
The sapwood is wide, pale pinkish-brown, and the heartwood is dark red or deep reddish-brown. It is a heavy, strong wood, with straight to somewhat interlocked grain and even and medium-fine texture, and weighing from 960 kg/m$^3$ to 1070 kg/m$^3$ when dried.

## Strength
Reported to be one of the strongest and hardest Indian timbers.

## Durability
Durable.

## Working qualities
Very difficult to saw owing to its hardness. It is said to machine moderately well, but is liable to tear up in rough streaks, if planed on a quarter-sawn surface.

## Uses
Its primary use is for heavy constructional purposes such as bridges and piling, but it is also used locally for keels, helms and masts for boats.

# PADAUK, ANDAMAN

*Pterocarpus dalbergioides*, Roxb.          Family: Leguminosae

## Other names
Andaman redwood, vermilion wood (USA).

## The tree
A large tree, up to 36m in height and 0.75m or more in diameter above the buttresses, which may be large. Above the buttresses

the stem is straight and cylindrical, and 12m or more to the first branch.

## The timber
The narrow sapwood is grey in colour, while the heartwood varies through shades of light yellowish-pink with slightly darker long light red lines, to brick-red, or deep purple with darker purple lines, presenting a beautiful roe figure and often a curly pattern on quarter-sawn surfaces. The grain is broadly interlocked, the texture is coarse, and the wood is moderately heavy, weighing about 790 kg/m$^3$ when dried.

## Drying
It air dries very well without warping and splitting, but it is liable to develop fine surface checks when converted green. Ring girdled trees reduce this tendency, but thick stock, particularly in wide sections may develop fine, curly, surface checks during air drying. It is said to kiln dry fairly easily without undue degrade.

## Strength
Similar to teak in all strength properties, but about 40 per cent harder.

## Durability
Very durable.

## Working qualities
The timber is not unduly difficult to saw and machine, and has only a moderate dulling effect on cutting edges, but it is rather difficult to obtain a good finish in planing and moulding due to the interlocked grain, particularly on quarter-sawn surfaces. A cutting angle of 15° is necessary, and provided the grain is adequately filled, the wood polishes well. It is a good turnery wood, holds screws well, but is difficult to nail.

## Uses
High-class joinery, furniture, bank counters, billiard tables, balustrades, flooring, and sliced, decorative veneer.

# PADAUK, BURMA

*Pterocarpus macrocarpus*, Kurz.          Family : Leguminosae

## Other names
pradoo, mai pradoo (Thailand).

## Distribution
Widely distributed in Burma and in the mixed deciduous forests of Thailand.

## The tree
A moderately large tree reaching a height of 25m and a diameter of 0.5m. The bole is commonly straight and cylindrical, but occasionally is crooked or forked. Straight boles are commonly about 8m long.

## The timber
The narrow sapwood is grey in colour, the heartwood varying from bright yellowish-red to dark brick-red, streaked with darker lines, but with age the colour dulls to a golden reddish-brown. The grain is interlocked in narrow bands, the texture is medium-coarse, and the wood weighs about 850 kg/m$^3$ when dried.

## Drying
Air dries reasonably well without excessive warping and splitting. There is a tendency for surface checking to occur in timber converted green, and for logs left for lengthy periods to develop cup-shakes unless the ends are protected. Ring girdling, and felling after the tree has died is said to give the best results. It kiln dries rather slowly, but quite well.

## Strength
The timber is very hard and strong, being some 40 per cent stronger in bending, resistance to shock loads, and hardness than Andaman padauk. It is also 30 per cent stronger in compression along the grain and 20 per cent stiffer.

## Durability
Very durable.

## Working qualities
Considered difficult to saw, especially when dry, and hard to work. It is generally easier to finish than Andaman padauk as the interlocked grain is not as pronounced; a cutting angle of 20° reduces the tendency for the grain to pick up during planing and moulding. It polishes well when the grain is properly filled, glues satisfactorily, but should be pre-bored if nailed.

## Uses
The value of Burma padauk is primarily due to its strength and durability, as opposed to Andaman padauk whose value is derived primarily from its decorative properties, accordingly, Burma padauk is used locally for the bottoms of railway trucks and waggons, wheel hubs, oil presses, and for constructional purposes. It is suitable for billiard table frames and sides, for furniture, and flooring for normal traffic, shop fitting, counters.

# PALI

*Palaquium ellipticum,* Benth.                Family : Sapotaceae

## Other names
panchonta, pala, kei pala.

## Distribution
Found in western India.

## The tree
A large tree with a fluted stem, varying in length from 5m to 12m and a diameter of 0.5m occasionally larger.

## The timber
The sapwood is pale red in colour and from 25mm to 50mm wide, while the heartwood is light red to light reddish-brown The wood is dull to somewhat lustrous, has a smooth feel, and weighs about 680 kg/m$^3$ when dried. The grain is usually straight, but is sometimes wavy, and the texture is medium and even.

## Drying
Dries well if care is taken otherwise liable to warp and split badly.

## Strength
Its strength properties are similar to those of teak.

## Durability
Moderately durable.

## Working qualities
Works and finishes relatively easily, but there is a tendency for some pieces to wind from the saw. It takes a good polish.

## Uses
Used in India for doors, window frames, planking, flooring, and for cheap furniture, and for shingles. Suitable-sized logs are in demand for masts and spars.

# PANAKKA

*Pleurostylia opposita*                    Family : Celastraceae
syn *P. wightii* W & A

## Other names
piyari.

## Distribution
Sri Lanka, southern India, Madagascar.

## General characteristics
The Celastraceae family consists of about 38 genera and more than 300 species, often shrubs and climbers, but including trees, whose timber is generally characterized by medium-sized to small, or extremely small vessels, bands of parenchyma, fine or very fine rays, arranged closely together, and a heartwood colour range from white, yellowish-white, grey, light red, rose-brown, yellowish-brown or light brown.
Many of the trees occurring in southern Asia, are small or medium-sized, and few are of economic significance except for relatively small decorative items, carving and combs.

*Pleurostylia opposita* is one such species, producing a timber light yellow-brick-red colour, with thin darker lines. It is a hard timber, very compact and firm, due to the minute vessels and

very fine rays, packed closely together. Wide bands of parenchyma, lighter in colour, cross the rays at regular intervals,
following the concentric layers of growth, wavy and broken.
It weighs about 768 kg/m³ when dry.

## Uses
Panakka has been used for many years in Sri Lanka for combs,
and more recently for mosaic flooring. Although fairly subdued
in appearance it is, nevertheless, a quite attractive wood, and
since it finishes very smoothly and is compact, it would seem
to be ideal for flooring of this type.

## *PTEROSPERMUM* SPECIES
Family: Sterculiaceae

### Other names
mayeng, welang, taung-petwun.

### Distribution
About 18 species of *Pterospermum* are distributed throughout
southern Asia, the Andamans, Sabah and the Philippines. At
least 11 species are found in India, 6 of which are found on the
west side of the peninsular.

### The tree
A fairly tall tree with a straight, cylindrical bole of some 9m to
12m in length and a diameter of 0.5m.

### The timber
The sapwood is whitish, and the heartwood of the various
species varies from yellowish with a red tinge, to a dull plum
colour. *P. canascens* syn *P. suberifolium* which occurs in
India and Sri Lanka, and is generally known as welang or
welanga, is light pinkish-red when freshly cut, turning an
agreeable darker red with a purple cast. The wood is fine textured
with a rather interlocked grain, and showing fine, inconspicuous
ripple marks. It weighs about 670 kg/m³ when dry.

### Drying
The timber dries extremely well and provided green stock is
protected from direct sunshine suffers little degrade, but there
is a tendency for rapidly dried wood to develop surface checks.

## Strength
No information.

## Durability
Moderately durable.

## Working qualities
Works easily with both hand and machine tools, and planes to an excellent finish. Can be stained and polished without difficulty.

## Uses
Furniture, panelling, high-class joinery and cabinet work, turnery, flooring.

# PUSSUR WOOD

*Carapa moluccensis,* Lam.                              Family : Meliaceae

## Other names
poshur, kyana.

## Distribution
Found in the coastal forests of Burma, the Andamans, and Bengal.

## The tree
It reaches its best development in Burma where it produces cylindrical and fairly straight boles some 6m in length and 0.5m in diameter.

## The timber
Similar to the associated crabwood or andiroba, of tropical America. (*C. guianensis*). The sapwood is narrow, brownish-white in colour, and the heartwood is red to deep wine-red or purplish red, often with darker bands or greenish streaks ; dull, with a greasy-smooth feel, straight to somewhat interlocked grain, and a fine and even texture. It weighs about 780 kg/m$^3$ when dried.

## Drying
Is reported to dry fairly well, and without undue degrade.

## Strength
No information.

## Durability
Moderately durable.

## Working qualities
It is said to work easily and well, and can be brought to a fine finish, taking a good polish.

## Uses
A high-class timber, distinctly ornamental, much in demand locally for chairs, cabinets, mouldings, boat fittings, tool handles and furniture generally.

# PYINKADO

*Xylia dolabriformis* Benth.       Family: Leguminosae
syn *Xylia xylocarpa* Taub.

## Other names
pyin, pran, pkhay, irul.

## Distribution
Pyinkado is widely distributed in Burma, where it occurs gregariously, and is also found in India.

## The tree
A fairly large tree, attaining a height of 30m to 36m and a diameter on favourable sites of 0.75m to 1.25m. In less favourable situations the stem is not so large or as well shaped, being sometimes irregular, occasionally fluted, and may be unsound in the centre.

## The timber
The narrow sapwood is pale reddish-white, and the heartwood is a uniform reddish-brown with few markings, or faintly veined with darker lines. It is a dull wood, with straight, wavy, or broadly interlocked grain, and medium texture. Gum deposits cause the wood to be speckled with dark tacky spots. The wood weighs on average 990 kg/m$^3$ when dry.

## Drying
The best results are obtained from green logs, when defects during air drying are usually minimal. Serious cracking in the log is generally confined to a few places, and surface checking in converted material is usually confined to short and wavy cracks. The timber dries slowly, and with care does not give much trouble.

## Strength
A strong, hard timber, the strength qualities being more marked in respect of crushing, hardness and shear, than to static bending. By comparison with greenheart, it is 15 per cent less hard, 20 per cent less resistant to suddenly applied loads, 25 per cent less stiff and weaker in compression along the grain, end 35 per cent weaker in bending.

## Durability
Very durable.

## Working qualities
Difficult to saw and machine, it has a severe blunting effect on cutting edges and a tendency to cause vibration in sawing and planing. Wavy and interlocked grain can be troublesome in planing and a reduced cutting angle is necessary to produce a smooth finish, but with care this can be obtained, and the wood is capable of taking a good and lasting polish.

## Uses
Pyinkado is eminently suited to heavy constructional work, especially where a high resistance to fungal and insect attack, crushing and abrasion is essential, as for example in bridge building, piling and for railway sleepers. It is well suited to flooring as a substitute for maple, and as a heavy duty decorative floor, and for wharf decking. It should be noted that plain-sawn material resists abrasion and wears more smoothly than quarter-sawn material, the former being best suited to flooring of the decorative type, the latter more desirable for decking and industrial flooring where a rougher surface is likely to be less slippery, especially in wet conditions.

# PYINMA

*Lagerstroemia speciosa* Pers.       Family : Lythraceae
syn *L. flos-reginae* Retz.

## Other names
jarul (India, Pakistan) ; banglang (Vietnam) ; intanin (Thailand).

## Distribution
Burma, Thailand, Vietnam and India.

## The tree
The tree attains a height of 24m and a diameter of 0.75m in India, but it grows to larger sizes in Burma where 12m boles with a diameter of 1.0m are common.

## The timber
The wide sapwood is greyish-white to pinkish-white, and the heartwood is light reddish-brown, rather lustrous and with a smooth feel. The wood is more or less straight-grained, and the texture is medium-coarse. It weighs about 640 kg/m³ when dried.

## Drying
Not difficult to dry ; it is best converted green and air dried in well-piled, open stacks. There is a tendency for surface checking to occur, but with care, this is not as a rule severe. It is liable to crack and split if left in the log.

## Strength
There appears to be a marked difference in strength between timber taken from the outside and from the centre of the heartwood, the latter being strong but stiff, while the former is more elastic. In general, pyinma is considered a moderately strong timber with strength properties 25 per cent inferior to those of teak.

## Durability
Moderately durable.

## Working qualities
Saws, works, and machines reasonably well, but material with

interlocked or irregular grain may give trouble in planing. It is capable of finishing to a smooth surface, and takes a good, lasting polish.

## Uses
Joinery, light construction, boat-building, turnery, and to a lesser extent for furniture.

The botanically allied *Lagerstroemia hypoleuca*, Kurz., produces Andaman pyinma, a slightly heavier wood (672 kg/m³) but similar in appearance to *L. speciosa*, coarser-textured, with larger pores and more widely spaced, narrower rays. Andaman pyinma is considered to be durable, and resistant to termites, and is used locally in building, for flooring, planking, door and window frames, good class furniture such as book cases, small tables, and for office equipment. The grain is usually much straighter than that of *L. speciosa*.

# ROSEWOOD, INDIAN

*Dalbergia latifolia*, Roxb.                    Family: Leguminosae

## Other names
Bombay blackwood, E. Indian rosewood.

## Distribution
Occurs throughout India, but is found at its best in southern India.

## The tree
A large tree with a cylindrical and fairly straight bole, averaging 6m in length, but much longer lengths may be found. The diameter varies according to locality and growth conditions from 0.3m to 1.5m.

## The timber
The sapwood is narrow, pale yellowish-white in colour, often with a purple tinge, and the heartwood ranges in colour from light, nearly golden-brown through shades of light rose-purple with darker streaks, to deep purple with rather blackish lines, darkening with age. The wood is dull, has a fragrant scent, and a

uniform and moderately coarse texture. The grain is interlocked in narrow bands, and the wood is heavy, weighing 870 kg/m³ when dried.

## Drying
Rosewood is probably the best of the hard Indian timbers to dry in log form or as hewn-square baulks prior to conversion. By this means colour is retained much better, and degrade is low, except for the heart centre which may shake badly and should be boxed out during conversion. The timber dries slowly. and should be protected against rapid drying both in the open air and in the kiln.

## Strength
It is a hard timber, and in comparison with English oak, is about 25 per cent stronger in bending and in compression along the grain, 15 per cent stiffer, and 20 per cent more resistant to suddenly applied loads.

## Durability
Very durable.

## Working qualities
Moderately hard to saw and machine, with a fairly severe dulling effect on cutting edges due to calcareous deposits. The reference to boxing out the heart during initial conversion (see Drying) is pertinent to sawing in final conversion; the large shakes which are liable to form in the centres of logs become filled with chalky deposits, as opposed to the smaller amounts which plug some of the vessels, and accordingly increase the tendency of the wood to blunt cutting edges. Planing and moulding produces a good finish when the cutting angle is reduced to 25°. The wood turns well, and with proper grain filling, takes an excellent polish, and an even better waxed finish. It produces really handsome decorative veneer.

## Uses
High-class furniture and cabinet work. In India it is also used for rammer heads, brake blocks, knee-timbers for boats, floors, and for house construction for posts, rafters.

# SAL

*Shorea robusta*, Gaertn. f.          Family : Dipterocarpaceae

## Other names
shal, sakwa, sala.

## Distribution
India and Bangladesh.

## The tree
A large, gregarious tree, with an erect, cylindrical stem attaining a height of 36m and an average diameter of 0.6m although the tree is capable of producing a diameter of 1.25m and more.

## The timber
The wood is typical of the balau group of *Shorea* species found in Malaysia, being pale brown in colour when freshly cut, turning dark reddish-brown after exposure, moderately heavy to heavy, with interlocked grain in broad bands, and an even, medium texture, and weighing between 800 and 880 kg/m$^3$ when dried.

## Drying
Dries slowly with a tendency to split, warp, and develop surface checks which usually follow an echelon pattern and is associated with the interlocking of the grain. With care in air drying, degrade should not be excessive.

## Strength
Similar to balau in general strength properties. By comparison with teak, sal is 30 per cent stronger in transverse strength, 100 per cent harder, and has a higher modulus of elasticity.

## Durability
Moderately durable.

## Working qualities
Moderately difficult to machine and work, with a moderate to severe dulling effect on cutting edges varying somewhat according to the degree of resin present in the wood. In planing and moulding there is a tendency for the grain to pick up,

especially on quarter-sawn surfaces, and a cutting angle of 20° is recommended. It takes nails and screws reasonably well.

## Uses
Heavy construction, bridges, piling, lock gates, waggon bottoms, hand-barrows, agricultural implements, dyeing vats, tent poles and pegs, and picking arms in textile mills.

Other species marketed as sal are as follows,
*Shorea obtusa*, Wall, occurs in Burma where it is known as thitya or Burma sal, and the botanically associated *Pentacme suavis* ADC (syn. *Shorea siamensis*, Miq.) known also as Burma sal or ingyin.

*S. obtusa* is a very heavy dark reddish-brown or brown wood, with interlocked grain, medium texture, resembling Indian sal but with broader rays. Weight 960 to 1070 kg/m³, locally considered very durable, and used for heavy construction, bridges, piling, etc.

*P. suavis* is a moderately heavy to heavy wood, weighing 850 kg/m³. It is a light yellowish-or-russet-brown timber, lacking the reddish cast that characterizes the woods of the balau group. It has an interlocked grain and a medium coarse texture. A durable wood used locally for bridges, piles and waggon construction.

## SANDALWOOD

*Santalum album,* Linn.                    Family : Santalaceae

### Other names
chandan, gandala, sukhad.

### Distribution
Generally found in open forests of central and southern India, and attaining its best growth on dry, stony, fertile soil.

### The tree
A small tree with an erect stem and a diameter of 0.3m on good sites, but otherwise smaller.

## The timber
The sapwood is whitish and unscented, and the heartwood is light yellowish-brown when freshly cut, turning to dark reddish-brown on exposure. The wood has an oily feel, and a strong odour of sandalwood oil which lasts for a very long time. The grain is straight or occasionally wavy, and very fine and even-textured. It weighs 960 kg/m$^3$ when dried.

## Drying
The wood dries slowly and easily without serious degrade.

## Strength
The wood is moderately hard to hard, and very close-grained, but strength is of no importance to the general use.

## Durability
Very durable.

## Working qualities
The wood works easily and well to a very smooth surface, and takes a satin-like finish, and is one of the finest carving woods.

## Uses
It has two primary uses, one for the extraction of the essential oil, the other for carving. It is used for small boxes, jewel cases, fan handles and combs.

# SATINWOOD, CEYLON

*Chloroxylon swietenia,* DC                    Family: Rutaceae

## Other names
E. Indian satinwood, satinwood.

## Distribution
Central and southern India and Sri Lanka.

## The tree
A small to moderate-sized tree, attaining its largest size in Sri Lanka. The bole is about 3m long, and the average diameter 0.3m or a little more.

## The timber
There is little distinction between sapwood and heartwood, the wood being light yellow or golden yellow in colour, with the inner wood a little darker than the outer. It has a satin lustre and a ribbon figure, sometimes broken or mottled. Gum rings may produce thin dark veins on longitudinal surfaces. The grain is narrowly interlocked, the texture is fine and even, and the wood weighs about 990 kg/m$^3$ when dried.

## Drying
Although the timber has a tendency to surface cracking, and a slighter tendency to warp and twist, careful treatment of material cut green during or immediately after the rains, should give good results.

## Strength
A very strong and hard timber with strength properties much higher than those of teak, but strength is of no real importance to the uses to which the timber is put.

## Durability
Durable.

## Working qualities
Owing to its density, it is a rather difficult timber to work with hand tools, and is fairly difficult to machine, having an appreciable dulling effect on cutting edges, with the grain tending to pick up during planing, particularly on quarter-sawn surfaces. This can be avoided by reducing the cutting angle to 15°. The wood turns well, and takes a fine and lasting polish.

## Uses
Furniture, cabinet-work, fancy goods, turnery, jute bobbins and decorative veneer.

## SIRIS, BLACK

*Albizia odoratissima* Benth.                     Family : Leguminosae

## Other names
thit-pyu (Burma) ; mara (Sri Lanka) *
*Mara timber presently on offer is described as *A. odoratissima*. This name was generally used as an alternative to describe kokko, (*A. lebbeck*). Both species are similar in appearance.

## Distribution
India, Pakistan, Bangladesh, the Andamans and Burma.

## The tree
A large, straight-stemmed tree with a cylindrical bole 9m or more clear of branches, and a diameter of about 0.6m.

## The timber
The timber is similar to kokko (*A. lebbeck*) in appearance and colour, but is somewhat finer-textured. The sapwood is wide, and whitish in colour, the heartwood brown to dark brown with lighter and darker streaks. The grain is shallowly interlocked and the texture is coarse. The wood weighs 688 kg/m³ when dry.

## Drying
Dries without serious splitting, and only slight warping. Green conversion and protection from hot sun during air drying is usually satisfactory.

## Strength
Strong and elastic, with strength properties similar to those of teak.

## Durability
Non-durable.

## Working qualities
Fairly difficult to work and machine, with a moderate to severe dulling effect on cutting edges. The grain is apt to chip and to pick up during planing, and a cutting angle of not more than 20° should be used. It takes a good polish when suitably filled.

## Uses
When properly worked and finished, the wood is quite attractive, and is used locally for decorative work and panelling, but it is also used for agricultural implements, dry cooperage and general construction.

# SIRIS, WHITE

*Albizia procera,* Benth                    Family : Leguminosae

## Other names
burda (Andamans) ; sibok (Burma).

## Distribution
India, Pakistan, Bangladesh, Burma and the North Andamans.

## The tree
A large tree with a straight stem producing boles 9m to 15m to the first branch, with a diameter of 0.5m or more.

## The timber
The sapwood is yellowish-white in colour, and the heartwood is brown, with lighter and darker bands, similar to kokko (*A. lebbeck*) in appearance, but less ornamental, lighter in weight and softer. The wood weighs on average 560 kg/m³ when dried, but this depends upon the amount of wide sapwood present; sapwood weighs 464 kg/m³ and heartwood weighs 640 kg/m³. The grain is straight or interlocked, and the texture very coarse but even.

## Drying
The timber air dries without difficulty if converted green and the sawn stock promptly stacked and protected from hot sun. Logs left for a time develop bad heart shakes. Kiln drying can be carried out without excessive degrade developing.

## Strength
The timber is strong, elastic, tough and hard. Compared to teak, it is 10 per cent stronger in modulus of elasticity, 25 per cent more resistant in compression parallel to the grain, and twice as hard.

## Durability
The heartwood is moderately durable.

## Working qualities
Moderately hard to work, and hard to saw by hand, but the wood planes to a smooth surface more readily than kokko and black

siris due to the less oblique grain angle. With careful filling, it takes a good finish.

## Uses
House construction, bridges, rice pounders, oars, and for furniture and table and counter tops.

# SISSOO

*Dalbergia sissoo,* Roxb.                          Family : Leguminosae

## Other names
shisham (Pakistan).

## Distribution
Sissoo grows naturally in India, Pakistan and Bangladesh, but also has been planted in irrigated plantations.

## The tree
In favourable localities the tree grows to 30m in height, with a 10m clear, cylindrical bole and a diameter of 1.0m. Elsewhere, the sizes are rather smaller.

## The timber
The sapwood is pale brownish-white, and the heartwood is golden brown to dark brown with deep brown streaks, similar in colour to Indian rosewood (*D. latifolia*), but lacking the characteristic odour of that wood, and generally more uneven and coarser-textured, with larger pores and wider rays. The grain is interlocked in narrow, straight lines, and the wood weighs about 830 kg/m$^3$ when dried.

## Drying
The timber is reported to dry slowly with little degrade. The best results were obtained by leaving the trees girdled for eighteen months before felling. Kiln drying is said to enhance the value of the timber by intensifying the difference in colour of the lighter and darker bands.

## Strength
No information.

## Durability
Very durable.

## Working qualities
It works and machines fairly easily, and gives a good finish. It turns well, and glues and polishes satisfactorily. It also peels well for veneer.

## Uses
High-class furniture, panelling, cabinet-making, boat building, plywood, and since the timber lends itself to bending, it is also used for bent-wood chairs and bent rims.


# TEAK

*Tectona grandis,* Linn. f.                    Family : Verbenaceae

## Other names
sagwan, teku, teka, kyun.

## Distribution
Indigenous throughout the greater part of Burma and the Indian Peninsular, in Thailand, Indonesia, etc. It has also been introduced into Malaysia, Borneo, the Philippines, and tropical Africa and Central America, etc, with varying success.

## The tree
Very variable in size according to local climatic and soil conditions. In favourable areas it is found with a long, straight, clean stem, but which, with advanced age becomes more fluted and buttressed towards the base, while in less favourable areas, particularly those with hot, dry, climates, the bole is more fluted, and the tree branches more. In some areas the bole is clear for 4.5m to 7.5m with a diameter of no more than 0.3m but in the more favourable locations such as the valleys of the larger rivers in Upper Burma, and in parts of the west coast of India clean boles may be 9m to 10m in length, with a diameter of 1.5m.

## The timber
The sapwood is usually narrow, pale yellowish-brown in colour, and the heartwood is dark golden-yellow darkening on

exposure to brown or dark brown, sometimes figured with dark markings. The wood is dull, with a rough, oily feel, and characteristically scented when fresh, the odour reminiscent of old leather. The grain varies from straight, typical of the best Burma teak, to wavy-grained, which is often more typical of Indian teak. The former material is generally more uniform in colour, with few markings, while the latter, especially from Malabar, is often more generously marked. The texture is rather coarse and uneven, and the wood weighs about 660 kg/m$^3$ when dried.

## Drying
Teak dries slowly but well, the main problem being the often erratic distribution of moisture throughout individual pieces, and between pieces.

## Strength
The timber is strong, moderately elastic, and hard, but is rather brittle along the grain. By comparison with English oak, air dry teak is some 10 to 20 per cent inferior in resistance to impact, but is stiffer and stronger in bending strength by a similar amount. Early tests carried out at Dehra Dun on naturally-grown and plantation-grown teak from Burma and Malabar indicated there were no essential differences in the general strength properties of either type.

## Durability
Very durable.

## Working qualities
Although the working qualities are somewhat variable, the wood in general works with moderate ease. The dulling effect on cutters varies from moderate to severe, but the wood is capable of a good finish if cutting edges are kept sharpened. It turns well, and can be glued, stained and polished. It holds screws and nails reasonably well, but the wood is inclined to be brittle. It produces veneer with ease.

## Uses
Teak enjoys a world-wide reputation owing to its durability, strength, ease of working, moderate weight, agreeable appearance, and small movement in fluctuating atmospheres. It is used for a vast number of purposes, for furniture, boat and

ship-building, for decking, deck-houses, rails, bulwarks, hatches, hulls, planking, oars and masts. It is used for high-class joinery, doors, staircases, panelling, veneer, plywood, dock and harbour work, piling, bridges, sea defences, flooring for light to moderate traffic, and for uses where chemical resistance is a requirement, eg laboratory benches and chemical vats. For the most decorative forms of furniture, the best marked teak is found in the drier regions of the Central Provinces, Deccan and the Malabar coast of India.

# THINGADU

*Parashorea stellata*, Kurz.                     Family: Dipterocarpaceae

**Other names**
kaunghmu, Tavoy wood.

**Distribution**
Found in Lower Burma.

**The tree**
A large tree with a long cylindrical bole, 18m or more long, and a diameter of 1.0m.

**The timber**
The wood is light yellowish-or-reddish-brown, with a rough feel. The grain is broadly and shallowly interlocked, the texture is coarse but even, and the wood weighs 710 kg/m$^3$ when dried.

**Drying**
The wood dries slowly with a tendency to warp and twist; it must be protected from full exposure to hot sun during air drying, and should be regarded as a moderately refractory species.

**Strength**
No strength data are available.

**Durability**
Probably moderately durable.

## Working qualities
The timber saws and works fairly well, but care is needed in planing to obtain a smooth surface. Can be polished after suitable grain filling.

## Uses
Boat-building, flooring and general construction.

# THITKA

*Pentace burmanica* Kurz.                      Family: Tiliaceae

## Other names
kashit (Burma).

## Distribution
Found in the tropical forests of Burma, mainly in Martaban, Pegu Yomas and Tenasserim.

## The tree
A tall tree with a clean, cylindrical, but sometimes curved stem, 0.5m in diameter, but on favourable sites such as in Tavoy, much larger sizes can be found. The bole is generally about 9m in length.

## The timber
The timber is sometimes called Burma mahogany because of its likeness to Cuban mahogany, but thitka is in no way related, and is a handsome species in its own right.

The sapwood and heartwood are not differentiated by colour, the wood is light red to terra-cotta, darkening on exposure, lustrous, with a typically interlocked grain, which produces a narrow regular stripe or roe figure on quarter-sawn surfaces, not unlike that of sapele. The wood contains conspicuous ripple marks, has a rather coarse texture, and weighs about 690 kg/m$^3$ when dried.

## Drying
Dries slowly with little degrade, but is liable to surface checking and warping unless care is taken.

## Strength
In the air dry condition, thitka is a harder, stronger wood than American mahogany (*Swietenia* spp.), being some 15 per cent stronger in compression along the grain, and about 30 per cent superior in resistance to shock loads. It is also 75 per cent harder than mahogany.

## Durability
Durable.

## Working qualities
Moderately easy to work and machine; it has a rather more dulling effect on cutting edges than does American mahogany, but it finishes cleanly, although the interlocked grain tends to pick up when quarter-sawn stock is planed or moulded. A reduction of cutting angle to 20° gives better results in these operations. It takes stain and polish very well, and has reasonable screwing and nailing properties.

## Uses
High-class furniture, mathematical instruments and instrument boxes, walking sticks, show-cases and general shop fitting, panelling, flooring, doors, and for gunwales, stringers, etc for boat-building.


# TULIP WOOD, BURMA

*Dalbergia oliveri,* Gamb.                     Family: Leguminosae

## Other names
tamalan (Burma).

## Distribution
Found in Burma, scattered in the teak and bamboo forests.

## The tree
A medium-size tree reaching a height of 24m and a diameter of 0.5m with a straight, cylindrical bole of about 7.5m.

## The timber
The narrow sapwood is yellowish-white, and the heartwood varies in colour from lemon-pink to red or reddish-brown with distinctly darker lines. The colour darkens on exposure. Straight, or slightly interlocked grain, and a medium-coarse texture, the wood is very hard and heavy, and weighs 1040 kg/m$^3$ when dried. The wood should not be confused with that of the tulip tree (*Liriodendron tulipifera*).

## Drying
The timber dries well, with only a slight tendency to split and surface check.

## Strength
No strength data are available, but the timber is said to be very strong, and fairly elastic.

## Durability
Very durable.

## Working qualities
Due to its density the timber is fairly difficult to work. When sharp cutting edges are maintained the wood can be worked to a beautiful finish. It takes a high and lasting polish. The wood also turns well.

## Uses
Turnery, parquet flooring, cabinet-making, decorative panels, bent-wood items, axe handles, brush backs.


# WALNUT

*Juglans regia* Linn.                    Family : Juglandaceae

## Other names
akrut, thitcha, akhor, khor.

## Distribution
Found in the Himalayas at elevations from 900m to 3000m from Afghanistan to Bhutan, and in the hills of northern Burma. It is also extensively cultivated.

### The tree
It can grow to very large sizes in the wild state, often reaching a height of 30m or more and a diameter of 2m, but more generally it has a clean, cylindrical stem of about 9m and a diameter of 1.0m.

### The timber
The sapwood is wide, and greyish-white in colour, and the heartwood is greyish-brown with few or no markings, or with darker streaks. The wood is variable in intensity of colour and markings, but beautifully mottled wood can be obtained by selection. The grain is straight, and the texture is medium and even. It weighs about 530 kg/m$^3$ when dried, rather lighter than the same species grown in Europe and Asia Minor.

### Drying
Dries slowly, and under local conditions requires care in air drying in order to avoid deep, fine splitting. Warping and surface checking is generally slight. The timber kiln dries without undue difficulty, but honeycombing may occur in thick material if the drying is forced.

### Strength
No information.

### Durability
Moderately durable.

### Working qualities
It saws easily, and lends itself to a marked degree to carving. It finishes to a smooth shiny surface and takes a high polish, and can be glued satisfactorily.

### Uses
Furniture, rifle stocks, carving, panelling and decorative veneer.

# PART II  SOFTWOODS

## CEDAR

*Cedrus deodara,* Louden                    Family : Pinaceae

### Other names
deodar, diar, dadar.

### Distribution
Found in the Himalayas from Afghanistan eastwards, at elevations of 1200m to 3000m, but more frequently at 1800m to 2700m. There are three types of deodar forest, the first in the dry arid zone of the inner Himalayas, the second in the intermediate ranges and valleys, and the third on the outer ranges, where the full effect of the monsoon is felt, and where growth is faster.

### The tree
A very tall and large-sized tree, reaching 60m or more in height and averaging 0.75m in diameter, although very much larger specimens are found.

### The timber
The sapwood is generally narrow, and whitish in colour, and the heartwood is light yellowish-brown when freshly cut, becoming a uniform brown colour after exposure. The wood exhibits unevenly spaced brown lines on longitudinal surfaces caused by occasional to frequent tangential rows of traumatic resin canals. It is rather oily, with a strong, characteristic cedar smell, a straight and fairly even grain, and a medium fine texture, and weighs about 580 kg/m³ when dried.

### Drying
Dries well, with little tendency to split or check, but with a slight tendency to warp.

### Strength
A reasonably strong timber similar to European redwood (*Pinus sylvestris*) in bending strength and stiffness, but slightly

heavier, and appreciably harder, and rather weaker in shock resistance and toughness.

## Durability
Very durable.

## Working qualities
Easy to saw and machine, and can be finished to a reasonable but somewhat oily surface.

## Uses
Building construction, joinery, bridge building and sleepers.

# CHIR

*Pinus longifolia,* Roxb.          Family : Pinaceae

## Other names
chil, sala dhup, gula.

## Distribution
Chir is found in the Himalayas from Afghanistan through Kashmir, and eastwards to Bhutan.

## The tree
The tree varies in size according to locality, but generally reaches a height of 45m. The average clear stem is about 12m with a diameter of 0.6m.

## The timber
The sapwood is creamy-white, and the heartwood light reddish-brown, a typical, resinous hard long-leaf pine, with distinct growth rings delineated by a darker band of latewood, an abrupt transition from earlywood to latewood, and transverse resin canals. It is a medium coarse-textured wood, with fairly straight grain in good specimens, but frequently with very twisted grain particularly towards the centre of the log. The wood weighs about 560 kg/m³ when dried.

## Drying
Green log conversion gives the best results, with storage under cover. The wood has a distinct tendency to split and check unless protected from rapid drying conditions.

## Strength
Clean, straight-grained stock has general strength properties comparable with deodar (*Cedrus deodara*), but strength in all categories is considerably reduced in twisted-grain stock.

## Durability
Moderately durable.

## Working qualities
Difficult to work; the resinous nature of the wood clogs saw teeth and the hardness of dead knots quickly dulls the edges of cutters. Any reduction of cutting angle below 30° for the purpose of improvement of planed or moulded surfaces is likely to result in chipped cutters caused by contact with dead knots. The wood takes varnish quite well.

## Uses
Building construction for light loading eg rafters, door and window frames, boarding, cheap joinery, railway sleepers.

# CYPRESS, HIMALAYAN

*Cupressus torulosa*, Don.            Family: Cupressaceae

## Other names
devidar, gellu, leuri.

## Distribution
Found in the Punjab and eastwards to Nepal, scattered, and often isolated, at 1800m to 2700m elevations.

## The tree
A very large tree, reaching an average height of 42m and 1.0m or more in diameter.

## The timber
The sapwood is white or pale yellowish-white, and the heartwood light yellowish-brown to light brown. The grain is usually very straight and even, and the texture is medium fine. The wood is very smooth, with a fragrant cedar-like scent, and weighs about 480 kg/m$^3$ to 512 kg/m$^3$ when dried.

## Drying
Dries quite well, with only an occasional long, fine, straight split developing.

## Strength
Its general strength properties are lower than those for deodar (*Cedrus deodara*), being about 25 per cent lower in compression parallel to the grain, and in modulus of elasticity. It is also about 11 per cent lower in transverse strength.

## Durability
Durable.

## Working qualities
It saws, planes, and performs well in other machine operations Since it is not very resinous, it holds paint well, and can be stained and polished.

## Uses
Joinery, including door and window frames, ceilings, door panels, skirting boards, and to a lesser extent for flooring. It is used for external purposes for shutters, weather boards and porch posts.

# HIMALAYAN SILVER FIR

*Abies pindrow*, Spach                    Family: Pinaceae

## Other names
rewar, span, morinda, drewar.

## Distribution
Found in the outer Himalayas from Afghanistan to Nepal.

## The tree
A large, lofty tree with a long straight stem. It attains a height of 56m to 60m and a diameter of 0.75m to 1.5m.

## The timber
The wood is whitish when freshly cut, turning light brown in time, with a mottling along the grain which becomes accen-

tuated as the wood darkens. The growth rings are delineated by a somewhat darker band of latewood, and while the transition from earlywood to latewood is gradual, the former is usually rather spongy, making the wood medium-textured, but coarser than spruce. The grain is straight and fairly even, and the wood weighs 432 kg/m³ when dried.

## Drying
The wood dries easily and well without warping and twisting, but it can develop long, serious cracks, if not protected during air drying.

## Strength
As a beam the timber is 10 per cent weaker than deodar, but 15 per cent stronger than spruce. In general it is considered to be somewhat stronger than the spruces and firs of North America.

## Durability
Non-durable.

## Working qualities
In its working and machining properties it is similar to European spruce; loosened knots can be a problem, and sharp cutting edges must be maintained in order to obtain a good finish. It stains effectively, and nails and glues satisfactorily.

## Uses
Flooring, boxes, water troughs, interior light construction.

## HIMALAYAN SPRUCE

*Picea morinda,* Link.                    Family: Pinaceae

## Other names
kachal, morinda.

## Distribution
Found from Afghanistan eastwards at elevations of 1800m to 3000m.

### The tree
A very large and tall tree reaching a height of 45m with a clear bole some 21m in length and a diameter of 1.0m.

### The timber
There is no distinction between sapwood and heartwood except that in large trees the wood near the centre and the butt may be a dull red colour, otherwise the wood is white when first cut, turning brownish with age. The grain is straight and even, and the texture medium-fine. It weighs about 465 kg/m$^3$ when dried.

### Drying
Dries easily and well, with only a slight tendency to warping and a definite tendency to splitting if not protected from hot sun.

### Strength
Similar to European spruce.

### Durability
Non-durable.

### Working qualities
Works easily and well and is similar in all respects to the working properties of European spruce, but the timber as used locally generally contains more and larger knots, and is therefore more difficult to plane or mould. It takes paint and varnish quite well.

### Uses
Structural usage in housing and for floors and skirtings, packing cases, water troughs.

## JUNIPER, INDIAN

*Juniperus macropoda*, Boiss.          Family: Cupressaceae

### Other names
appurz, shupa, dhupri, dhupi.

### Distribution
Found in the inner Himalayas from Afghanistan to Nepal.

## The tree
A generally large though not very tall tree, the stem is often fluted, with low branches contributing to very knotty timber. The diameter is about 0.5m on average, but much larger dimensions are recorded.

## The timber
The sapwood is thin, and yellowish in colour, and the heartwood dull red to reddish-brown, often with a purplish cast. It has a straight grain, and a fine and even texture, but is subject to knots. It has a fragrant cedar-like smell, but is non-resinous. Weight about 450 kg/m³ when dried.

## Drying
It dries slowly and well, neither warping nor splitting, but knotty material is liable to checking in the vicinity of knots.

## Strength
No information.

## Durability
Probably durable.

## Working qualities
Knot-free material works and machines very easily, finishing excellently, but the disturbed grain around knots tends to tear. It is inclined to split when nailed and requires care in screwing.

## Uses
Good clean material is used for pencil slats, but otherwise the timber is employed locally for building.

## PINE, BLUE

*Pinus wallichiana*                     Family : Pinaceae
syn *Pinus excelsa,* Wall.

## Other names
biar, kail, chil, chila.

## Distribution
Found from Afghanistan to Sikkim and Bhutan, sometimes in pure stands, sometimes in association with deodar.

## The tree
A large tree, attaining a height of 30m to 36m with a straight, cylindrical stem some 18m in length and a diameter commonly of 0.6m but occasionally much more.

## The timber
The sapwood is whitish, and the heartwood light pinkish-red to light red, fairly lustrous, with a straight and fairly even grain and a medium-fine texture. A soft pine similar to yellow pine (*P. strobus*) but coarser grained because of the more pronounced latewood. Weight about 512 kg/m$^3$ when dried.

## Drying
Dries well, and considerably better than other Indian pines.

## Strength
A moderately strong timber, equal in shear strength to deodar, but 36 per cent weaker in transverse strength, and 16 per cent below that timber in compression parallel to the grain.

## Durability
Non-durable.

## Working qualities
Easy to saw and work, and can be planed and moulded to a smooth surface provided there are not too many knots present. Takes paint, varnish and polish quite well but is rather absorbent.

## Uses
Construction, joinery, flooring, decking for boats, and when selected, for drawing boards.

# USE GUIDE FOR SOUTHERN ASIAN TIMBERS

## ACID AND DYE VATS
sal, teak

## BOAT AND SHIP CONSTRUCTION

### Decking
aini
amari
cedar, white

gumhar
pine, blue
teak

### Framing
bow wood

champak

### Keels and stems
amoora
babul (heartwood)

mesua
pyinma

### Masts and spars
aini
benteak
'Burmese cedar'
chaplash

chuglam, white
mesua
pali
teak

### Oars
amari
babul (heartwood)
chaplash
chuglam, white
didu

gumhar
hollock
siris, white
teak

### Planking
amari, teak

### Superstructures
aini
amari
amoora
'Burmese cedar'
chuglam, white

padauk
pyinma
teak
thitka
walnut

# BOXES AND PACKING CASES

'Burmese cedar'
canarium, Indian
cedar (deodar)
chaplash
chuglam, white
didu
eng
fir, silver

grandis gum
gurjun
haldu
kindal
lunumidella
pine, blue
spruce

# CONSTRUCTION

## Heavy
anan
anjan
axlewood
bullet wood
eng

gurjun
mesua
pyinkado
sal
yon

## Light
aini
benteak
bombay, white
'Burmese cedar'
canarium, Indian
cedar (deodar)
cedar, white
champak
chaplash
chuglam, black
chuglam, white
chir
cinnamon
cypress
fir, silver

grandis gum
gumhar
haldu
hollock
juniper
kindal
kokko
padauk (Burma)
pine, blue
pyinma
siris, black
siris, white
spruce
thingadu

# DOORS

aini
'Burmese cedar'
cedar, white

champak
chuglam, black
chuglam, white

340

## DOORS *(continued)*

cinnamon
cypress
gumhar
pali

teak
thitka
walnut

## FANCY GOODS

blackwood, Burma
ebony
pussur wood

marblewood
sandalwood
satinwood

## FLOORING

canarium, Indian
cedar, white
chuglam, black
chuglam, white
cypress
eng
grandis gum
gurjun
haldu
kindal
padauk

pali
panakka
*Pterospermum* spp.
pyinma
pyinkado
rosewood, Indian
spruce
teak
thingadu
tulipwood, Burma

## FURNITURE AND CABINET MAKING

aini
amari
blackwood, Burma
bombway, white
boxwood (inlay)
bullet wood
'Burmese cedar'
cedar, white
champak
chickrassy
chuglam, black
chuglam, white
dhaman

ebony
elm, Indian
gumhar
haldu
kokko
laurel, Indian
marblewood, Andaman
padauk, Andaman
*Pterospermum* spp.
pussur wood
rosewood, Indian
satinwood
siris, black

siris, white
sissoo
teak

thitka
tulipwood, Burma

## JOINERY

**High-class**
aini
amari
'Burmese cedar'
cedar, white
chickrassy
chuglam, black
chuglam, white
gumhar
haldu

kokko
laurel, Indian
padauk, Andaman
*Pterospermum* spp.
siris, black
teak
thitka
tulipwood, Burma
walnut

**Utility**
amoora
babul
benteak
bombway, white
canarium, Indian
cedar (deodar)
cedar, white
cinnamon
cutch
cypress
kindal
pali

pine, blue
pyinma
semul
champak
chaplash
chir
elm, Indian
fir, silver
hollock
siris, white
spruce
thingadu

## MARINE PILING AND CONSTRUCTION

**Under water**

**(a) Teredo infested waters**
pyinkado, teak*

*tests carried out in Burma showed these two timbers to have a reasonable resistance to Teredo attack, while Indian laurel and sal had a fair resistance. In these experiments both green-heart and jarrah were tested, and both were severely attacked.

## MARINE PILING AND CONSTRUCTION *(continued)*

**(b) Non-teredo waters, in addition to above,**

anan
bullet wood
Indian laurel
mesua

palu
pali
sal

## Above water

**(a) Docks, wharves, bridges, etc**

anan
anjan
axlewood
bullet wood
cedar (deodar)
gurjun
mesua

pali
palu
pyinkado
sal
siris, white
teak

**(b) Decking**

axlewood
cedar, white
chuglam, black
gurjun

pali
pyinkado
sal
teak

## MUSICAL INSTRUMENTS

boxwood
'Burmese cedar'
cutch
chuglam, black
ebony
gumhar

haldu
laurel, Indian
lunumidella
padauk, Andaman
rosewood, Indian
thitka

## SPORTS GOODS

benteak
boxwood
'Burmese cedar'
chuglam, black
ebony

lilac, Persian
padauk, Andaman
padauk, Burma
walnut

343

## STAIR TREADS

| | |
|---|---|
| chickrassy | haldu |
| chuglam, black | kindal |
| chuglam, white | |

## TERMITE RESISTANCE (HEARTWOOD) *

### Very resistant

| | |
|---|---|
| aini | padauk, Burma |
| anan | pali |
| anyan | Persian lilac |
| babul | pyinkado |
| bullet wood | rosewood |
| Burma blackwood | satinwood |
| cinnamon | sissoo |
| cutch | teak |
| kindal | white cedar |
| mesua | |

### Resistant

| | |
|---|---|
| 'Burmese cedar' | gumhar |
| cedar (deodar) | Indian laurel |
| chaplash | padauk, Andaman |
| chuglam, black | pussur wood |
| cypress | pyinma (Andaman) |
| dhaman | thitka |

### Moderately resistant

| | |
|---|---|
| amari | kokko |
| amoora | pali |
| benteak | sal |
| black siris | walnut |

### Susceptible

| | |
|---|---|
| axlewood | chickrassy |
| babul (sapwood) | chir |
| blue pine | chuglam, white |
| bombway, white | didu |
| bombax, Indian | elm |
| canarium, Indian | eng |
| champak | gurjun |

344

**Susceptible** *(continued)*

haldu
hollock
juniper
lunumidella
pyinma (Burma)

silver fir
spruce
white siris
yon

*The above classification refers to resistance to attack by both subterranean and dry-wood termites. Where the resistance to either type differs, the lower rating is given.

## TURNERY

babul
boxwood
bullet wood
cutch
ebony
gumhar
haldu
laurel, Indian
lilac, Persian

marblewood, Andaman
*Pterosphrmum* spp.
pyinma
rosewood, Indian
sandalwood
satinwood
sissoo
tulipwood, Burma

## VEHICLE BODIES

aini
anan
babul
benteak
cedar, white
eng
gumhar

gurjun
haldu
kindal
laurel, Indian
pyinkado
sal
siris, white

## VENEER AND PLYWOOD

**Corestock**

bombax, Indian
'Burmese cedar'
champak

didu
hollock

**Decorative**

amari
'Burmese cedar' (moderately)

chickrassy
laurel, Indian

**Decorative** *(continued)*

lilac, Persian
padauk, Andaman
rosewood, Indian
satinwood

sissoo
teak
walnut

**Utility (plywood, chip-baskets, small laminated items etc)**

amari
bombax, Indian
bombway, white
'Burmese cedar'

champak
chaplash
hollock

# AMENABILITY OF HEARTWOOD TO PRESERVATIVE TREATMENT

**Extremely resistant**

anjan
babul
benteak
blackwood, Burma
bullet wood
'Burmese cedar'
canarium, Indian
chaplash
cutch

kindal
marblewood
mesua
padauk, Burma
pali
pyinkado
siris, black
siris, white
teak

**Resistant**

aini
anan
axlewood
elm, Indian

eng
gurjun
sal
thitka

**Moderately resistant**

amoora
bombway, white
cedar, white
champak
chickrassy
chuglam, black
chuglam, white
dhaman
fir, silver
gumhar

haldu
hollock
juniper
laurel, Indian
lunumidella
padauk, Andaman
pyinma
spruce
thingadu

**Permeable**

| | |
|---|---|
| bombax, Indian | cypress |
| cedar (deodar) | pine, blue |
| chir | yon |

## AMENABILITY OF HEARTWOOD TO PRESERVATIVE TREATMENT

The above classification refers to the ease with which a timber absorbs preservative under both open-tank (non-pressure) and pressure treatments. Sapwood, although nearly always perishable, is usually more permeable than heartwood, accordingly, the above classification refers to the relative resistance of heartwood to penetration.

### Extremely resistant
Timbers that absorb only a small amount of preservative even under long pressure treatments. They cannot be penetrated to an appreciable depth laterally, and only to a very small extent longitudinally.

### Resistant
Timbers difficult to impregnate under pressure and require a long period of treatment. It is often difficult to penetrate them laterally more than about 3mm to 6mm.
Incising is often used to obtain better treatment.

### Moderately resistant
Timbers that are fairly easy to treat, and it is usually possible to obtain a lateral penetration of the order of 6mm to 18mm in about 2-3 hours under pressure, or a penetration of a large proportion of the vessels.

### Permeable
Timbers that can be penetrated completely under pressure without difficulty, and can usually be heavily impregnated by the open-tank process.

# 4

# SOUTH EAST ASIA

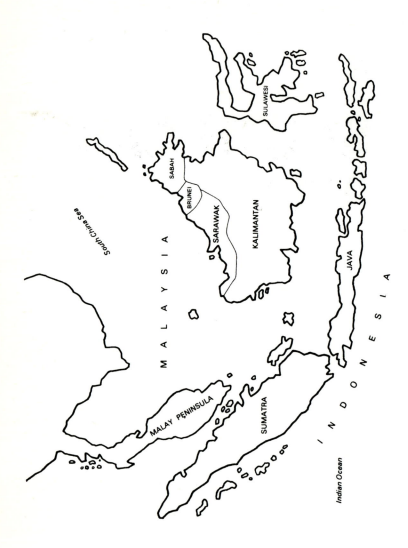

# INTRODUCTION

The geographical scope of this chapter is confined to the areas of south-east Asia known as Malaysia and Indonesia, and accordingly it embraces mainly the Malay peninsula, Borneo (Kalimantan, Sabah, Sarawak, Brunei etc), Sumatra and Java. It was considered necessary to confine the coverage of important timber species in this way because the forest areas of the entire continent of Asia fall into three fairly well-defined groups. Some 42 per cent of the forests are coniferous and occur in the Himalayas and northwards, and in the mountains of Asia Minor, China and Japan. About 28 per cent are of temperate hardwoods or are mixed forests, and occur in similar areas while tropical hardwoods, making up some 30 per cent of the forests, occur south of the Himalayas and in many countries comprise 100 per cent of the woody species.

As in tropical forests elsewhere the number of species is very great, and in order to give as wide a coverage as possible to the present and potential timbers, and since Malaysia and Indonesia form a separate source of commercial timber exploitation largely distinct from South Asia, this chapter is concerned mainly with the relatively limited area of south-east Asia. It does not follow that all the timber species described are indigenous only to the areas mentioned, since nature is not governed by political boundaries, and in some cases the species may be found elsewhere. For a wider appreciation of the timber resources, reference should be made also to the previous chapter on timbers of Southern Asia.

## Malaysia

Malaysia is a Federation of thirteen states and comprises two regions, the Malay peninsula or West Malaysia, and the Borneo states of Sarawak and Sabah, or East Malaysia. Throughout Malaysia as a whole, some 20917 kilometres of first class surfaced roads exist, and more than 7079 kilometres of railway tracks are in service. The main shipping ports are at Kelang and Penang in West Malaysia, and at Kota Kinabalu in East Malaysia. Bahasa Malaysia (Malay language) is the national and official language, but English is widely spoken in commerce and industry.

The long, narrow peninsula of West Malaysia extends from the isthmus of Kra to Singapore, a distance of some 1200 kilometres and its maximum width is about 290 kilometres. The backbone of the peninsula is a system of forested mountains which yield many fine timbers. The coastal areas are usually swampy and mangrove-lined, but the needs of both ecology and economic progress are finely balanced, and while in present day West Malaysia more land is being opened up to agriculture, forestry is not neglected, and science and technology are helping to conserve the forests by the maintenance of forest reserves; and by judicious production and grading techniques, and by grouping similar generic species into commonly acceptable commercial timbers has developed a source of tropical hardwoods available to world markets of present and future potential.

Although Sarawak and Sabah form present day East Malaysia, they cannot be isolated from Borneo as a whole in the context of timber resources, particularly to the west and north of the island. The surface of Borneo is generally mountainous. There is a central group of mountains from which other ranges radiate, but the systems are confused. The island is rich in all kinds of tropical products, from numerous palms to valuable hardwoods, but generalizing, it is perhaps true to say that timber production is concentrated to the north and north-west of the island, and minerals, coal, and other main industries mainly in the west and south.

### Indonesia

For the purpose of this chapter Indonesia is taken mainly as the islands of Sumatra and Java.

Sumatra is separated from the Malay peninsula on the east by the Malacca Straight, and from Java on the south-east by the Sunda Strait. The west coast, fronting the Indian Ocean is formed by parallel ranges of mountains, and from them the ground slopes to the east coast in rich alluvial soil. The mountains are heavily forested and with the lower slopes provide many valuable timber species.. The main ports are Padang, Sabang, Palembang, Aru Bay and Tanjong Pinang.

Java is separated from Borneo by the Java Sea, from Sumatra on the west by the Sunda Strait, and from the island of Bali by the Bali Strait. It is low-lying and covered by mangrove swamps in the north, while the interior is mountainous. The south coast is bold and rugged, but the north coast is indented by numerous bays, affording shelter and commodious harbours, the chief of which are Djakarta and Surabaya. Large areas of the island are devoted to agriculture, *but there are extensive forests, producing valuable timbers, and in particular teak, which occurs on the lower slopes.*

*Throughout Malaysia and Indonesia many of the forests are dominated by a vast variety of tree species of the Dipterocarpaceae and Leguminosae familes, the former in particular representing a valuable contribution to the world's timber resources, and for that reason some effort has been made in the preparation of this chapter to include and tabulate the various species that are frequently employed in the production of a single commercial timber, and which often appear confusing. The various species of Shorea, for example, have been separated and segregated into the different groups applicable to the commercial nomenclature, and this is reflected in the index.*

# PART I HARDWOODS

## AMBOYNA

*Pterocarpus indicus* Willd.          Family: Leguminosae
and possibly allied species

### Other names
narra (Philippines, USA) ; padauk (Burma).
Various species of *Pterocarpus* occur in South East Asia, the
Andamans and the Moluccas, and are described elsewhere in
this booklet. The trade name amboyna is applied to certain
highly figured, reddish-brown wood obtained from burrs which
develop more especially in *P. indicus*, but may occur in allied
species.

### The timber
Amboyna varies from light brown with reddish markings to deep
rich blood-red with blackish markings. The sapwood is about
50mm wide, and white to pale straw colour. The heartwood is
usually marked with very small twisted curls and knots similar
to birds-eye maple, but more varied in effect. The texture is
medium but may be uneven due to grain disturbances, while
the grain is wavy to interlocked. The wood weighs about
670 kg/m³ when dried.

### Drying
Reported to dry well, with little shrinkage. Its character suggests
however that care is required in order to avoid surface checking
which could be a serious defect in expensive veneer.

### Durability
Very durable.

### Working qualities
Reported to be fairly easy to work, but some care is required in
planing quarter-sawn material as the irregular grain tends to
pick up during this operation. Takes a good finish, stains readily,
but may require filling before polishing. The timber turns well.

### Uses
Decorative veneer, high-class furniture and cabinet-making.

# BANGKIRAI

*Shorea laevifolia* (Parijs) Endert.     Family : Dipterocarpaceae

Of the many species of *Shorea* growing throughout Malaysia and Indonesia, *Shorea laevifolia* is confined mainly to East Kalimantan, from whence is shipped commercial bangkirai. The timber approximates to the balau group of heavy hardwoods shipped from Malayasia, but whereas balau is produced from a number of species of *Shorea*, bangkirai is the product of a single species.

## General characteristics
There is little distinction by colour between sapwood and heartwood which is a pale golden-brown. The grain is interlocked, and the texture is medium to fine. The weight varies from 800 to 980 kg/m³ when dried, the lighter weight occurring in the centre of large, over-mature logs.
The wood dries without serious degrade, but there is a distinct tendency for surface checks and end splitting to develop, and a slight tendency for bowing and twisting to occur.

## Strength
Bangkirai compares favourably with jarrah, but is slightly more elastic.

## Durability
Durable to very durable.

## Working qualities
Although hard and heavy, the timber works reasonably well but with fairly severe blunting of normal cutting edges. A smooth surface can be obtained in planing and moulding when cutting angles are reduced to 20°.

## Uses
Bangkirai is used extensively for marine construction, lock gates, piling, fenders, bridge components, decking and walkways, heavy-duty industrial flooring, strip and parquet flooring, carillon framing in church towers, oil rig construction, sills and thresholds, etc. It has a high resistance to acids and is suitable for floors in chemical works, acid and dye vats, and, despite its

355

hardness has been used in Singapore for sliced veneers. With regard to the lighter-weight centres of some large logs, it is possible to produce 'boxed hearts' in certain dimensions.

# BELIAN

*Eusideroxylon zwageri* Teijasm and Binn.                    Family: Lauraceae

### Other names
tambulian (Sabah); billian, boelian, borneosch ijzerhout, onglen (Indonesia); Borneo ironwood (UK).

### Distribution
It occurs in East Kalimantan, Sabah, Brunei and Indonesia. It is fairly abundant on the east coast of Sabah, and may occur up to 150m above sea level. It is occasionally gregarious, tending to grow in clusters.

### The tree
A medium sized tree, attaining a height of 30m and a diameter of about 0.5m.

### The timber
The sapwood is yellowish in colour, turning darker on exposure, and is usually distinct from the heartwood which is yellowish-brown to reddish-brown when freshly cut, sometimes with a greenish tinge, but darkening considerably on exposure to a deep brown. The grain is usually straight but may be slightly interlocked, and the texture is moderately fine and even. The wood is exceptionally hard and heavy, weighing about 1040 kg/m$^3$ when dried.

### Drying
Should be dried slowly in order to avoid surface checking, splitting and warping.

### Strength
A very strong, tough timber, about 50 per cent stronger in static bending and stiffness than English oak.

## Durability
Very durable.

## Working qualities
Due to its hardness and toughness, it is a difficult timber to work. Cutting edges are quickly blunted and require frequent sharpening. With care, the timber can be finished to a smooth surface.

## Uses
Marine piling, bridge construction, wharves and decking, shingles, flooring, printing blocks, ship building, and for other uses where high durability and strength are required.

# BINTANGOR

*Calophyllum* spp.　　　　　　　　　　　Family: Guttiferae
excluding *C. inophyllum* L.

## Other names
None recorded.

## Distribution
The genus *Calophyllum* is widely distributed throughout South East Asia, where its habitat ranges from coastal and swamp districts to mountain forests. Malaysia is the main source of supply.

## The tree
The trees of the various species are variable in size, but generally they reach a height of 30m and a diameter of 0.6m, and although variable in form according to environmental conditions the trees are often straight and slender, with a clear bole of 15m to 18m.

## The timber
The sapwood varies from yellowish-brown to orange, and is usually distinct from the heartwood which is pinkish or reddish-brown, darkening on exposure to brick red or dark orange. Planed surfaces show a high lustre, and a stripe figure occurs on quarter-sawn surfaces, while tangential surfaces show a dark

coloured zig-zag marking. The texture is rather coarse and uneven, and the grain interlocked, spiral or wavy. The wood is moderately hard and heavy, the weight varying from 560 kg/m³ to 768 kg/m³ but generally averaging 690 kg/m³ when dried.

## Drying
The timber requires care in drying due to its tendency to warp and twist. Weighting down of the stack during air drying will reduce this tendency, and a mild schedule is recommended in kiln drying.

## Strength
The strength varies according to species, but on the whole the average strength compares with European birch.

## Durability
Moderately durable.

## Working qualities
Fairly easy to saw and work, but sawn surfaces tend to be woolly. It can be planed to a fairly satisfactory smooth surface, but care is needed on quarter-sawn surfaces. Takes a good finish but requires filling before polishing.

## Uses
Bintangor is used locally for masts and spars, boat-building, flooring, furniture, cabinet work, bridge-building, and general construction, and it has also proved excellent for diving boards.


# BINUANG

*Octomeles sumatrana* Miq.　　　　　　Family: Datiscaceae

## Other names
erima, ilimo (New Guinea).

## General characteristics
*Octomeles sumatrana* occurs in Sabah, Sarawak, Indonesia and Papua New Guinea. It is a soft, light-weight timber, light-straw in colour with no distinction between sapwood and heart-

358

wood. It weighs about 400 kg/m³ when dried. It is perishable but permeable, and is used for packing cases and light interior construction and carpentry.

# BITIS

### Species of Sapotaceae

Bitis is the trade name in Malaysia for a few species of the Sapotaceae family which produce heavy hardwoods. The bulk of commercial shipments is comprised mainly of the timber produced from *Madhuca utilis*, but certain species of *Palaquium* may be included, in particular, *P. ridleyi*, and *P. stellatum*.

### General characteristics
The sapwood is well defined from the heartwood which is reddish-brown or chocolate red-brown in colour, with an occasional stripe figure on quarter-sawn surfaces. Planed surfaces have a medium lustre, and the wood is fairly straight-grained, with a moderately fine and even texture. The wood is very hard and heavy, weighing from 930 kg/m³ to 1120 kg/m³ when dried.

### Drying
The wood is difficult to dry. It gives up its moisture slowly, there is a distinct tendency to surface checking, and shrinkage rates are high.

### Strength
Bitis is a heavy and strong timber, with strength properties in the greenheart class or better.

### Durability
Very durable.

### Working qualities
Difficult to work, particularly with hand tools. The hard, horny nature of the wood blunts saws and cutters fairly rapidly, but the compactness of the wood generally produces a smooth surface in planing. It is reputed to take stain and polish satisfactorily.

### Uses
Heavy constructional work of the most exacting nature, paving blocks, agricultural implements and turnery.

# CANARIUM, MALAYSIAN

Species of *Canarium* and other genera        Family: Burseraceae

**Other names**
kedondong (Malaya).
The genus *Canarium* is found in many tropical areas of the world, but the timbers produced from the various species although somewhat similar in appearance have previously been known under a variety of local trade names, often confusing. In order to bring these various descriptions into a common basis, BSI has recommended the standard name canarium for all these woods, the description being qualified by the country of origin, eg Indian canarium, African canarium, and Malaysian canarium. Insofar as India and Africa are concerned the timber is more generally produced exclusively from species of the genus *Canarium*, but that produced in Malaysia under the local name of kedondong often includes species of *Dacryodes* and *Santiria* as well as *Canarium*. The major species involved are as follows, those underlined being more generally included in commercial kedondong.
*Canarium littorale*, *C. patentinervium*, *C. psuedodecumanum*.
*Dacryodes costata*, *D. rostrata*, *D. rugosa*, *D. laxa*, *D. macrocarpa*, *D. puberula*, *D. rubiginosa*.
*Santiria laevigata*, *S. tomentosa*, *S. nana*, *S. rubiginosa*.
All these species are closely related botanically, and their timber is similar in appearance and properties.

**The tree**
The trees are moderately large, reaching a height of 30m and a diameter of 0.6m or a little more. The boles are usually clear for about 12m above the small buttresses.

**The timber**
The sapwood is lighter-coloured than the heartwood but the boundary is not clearly demarcated. The heartwood is light pinkish-brown or reddish-brown, sometimes with a pale yellowish tinge. According to species, planed surfaces show a variable lustre, ranging from rather dull to very lustrous. The grain is often wavy, and is shallowly interlocked, and the texture is moderately coarse and even. The average weight is 580 kg/m³ when dried.

## Drying

Fairly easy to dry, but there is a tendency for end splits to develop and existing shakes to extend. Shrinkage is reported to be low, and of the order of 3.5 to 5.0 per cent tangentially, and 2.0 to 3.0 per cent radially, in drying from the green state to oven dry.

## Strength

Similar to red meranti.

## Durability

Non-durable.

## Working qualities

Relatively easy to work, but with moderate to severe blunting of cutting edges. Quarter-sawn material tends to pick up in planing, but reduction of the cutting angle to 20° is usually satisfactory in providing a smooth surface. Nails, glues and screws well, and is suitable for veneer.

## Uses

Veneer, boxes, crates, joinery and light construction.

# CHENGAL

*Balanocarpus heimii* King               Family : Dipterocarpaceae

Chengal is found throughout Malaysia, and is the product of a single species of the genus *Balanocarpus*. The following is a description of the Malaysian chengal.

## General characteristics

The sapwood is well defined from the heartwood which is yellowish-green when freshly cut, turning to dark tan-brown on exposure. Planed surfaces are lustrous, often with a vague stripe figure. The grain is typically interlocked, and the texture fine and even. Storied rays and elements produce characteristic and very distinct ripple marks on tangential surfaces. The wood is hard and heavy, and weighs about 930 kg/m$^3$ when dried. Chengal is reported to dry relatively easily and well, and to remain stable in service.

## Strength
Chengal has strength properties similar to those of greenheart, but is slightly inferior to that timber in compression and shear strength.

## Working qualities
Fairly difficult to work because of its hardness and interlocked grain, but it is capable of being worked to a smooth finish.

## Uses
Chengal is a very durable and strong timber which makes it suitable for the most exacting uses. It is used for all types of heavy construction, flooring, decking, boat building, sleepers, and for handles, levers, poles and posts.

# DURIAN

## Species of Bombacaceae

The Bombacaceae family includes many tree species which are valuable sources of fibres and fruits, but which are in some cases also valuable for their timber. The family includes *Bombax malabaricum* (Indian cotton tree), *Ceiba pentandra* etc, (ceiba, or silk cotton tree of the tropics), and *Ochroma pyramidale* (balsa, or polak of tropical America). Those timbers produced by the Bombacaceae are either very light in weight and soft, or relatively light-weight, and relatively weak and soft.

The timber commercially known as durian takes its name from the durian fruit of Malaya, a large, conical spiny fruit some 150mm in diameter, much sought after by the native Malaysians. The main source of this fruit is *Durio zibethinus* L., a native of Malaysia, but occurring in Burma, and the East Indies generally.

Durian timber from Malaysia is produced not only by this species but also by several closely related species, as given below, those which dominate commercial supplies being underlined.

*Durio zibethinus* L., *D. oxleyanus* Griff., *D. lowianus* Scort., *D. oblongus* Mast.

*D. griffithii*, and *D. graveolens. Coelostegia griffithii*, and *Neesia altissima, N. kostermansiana, N. malayana*, and *N. synandra*.

### General characteristics
All the above species (except *Coelostegia griffithii*), have pinkish-brown, red-brown, or deep red-brown heartwood, the sapwood being a lighter colour, and distinct. *C. griffithii* has orange-brown or orange-red heartwood with little distinction from the sapwood. The wood is plain, without lustre or figure, except that the larger rays produce a fleck on radial surfaces. The grain is slightly interlocked, and the texture coarse and uneven, and the wood weighs on average 688 kg/m$^3$ when dried.

There is no information regarding drying and strength properties, but the timber is said to work easily, but is difficult to obtain a smooth finish. It is non-durable, and is used locally for packing cases, tea chests, toys, and for interior construction.

## GERONGGANG

*Cratoxylon arborescens* (Vahl.) Bl.          Family: Guttiferae
and *C. arborescens* var. *miquelli* King

### General characteristics
Geronggang is a light Malaysian hardwood, the sapwood of which is pink and distinct from the heartwood which is brick-red or deep pink, with a very lustrous surface when planed. The grain is interlocked, and the texture moderately coarse but even. It is a soft wood to cut, light in weight, ranging from 432 kg/m$^3$ to 608 kg/m$^3$ but commercial parcels usually average about 544 kg/m$^3$ when dried. The wood is easy to dry, relatively easy to work, and takes a good finish. It is non-durable.

### Uses
A general utility wood, suitable for interior joinery, panelling and packing cases.

## GIAM

*Hopea* spp.                    Family: Dipterocarpaceae

### Other names
selangan batu (Sabah).
**Giam** is the group trade name in Malaysia and Sarawak, for certain heavy species of *Hopea* formerly grouped with certain

363

heavy species of *Shorea* under the description, **selengan batu**. Although Sabah may still retain this description, the timber is perhaps now better known as **giam**. **Selangan batu** is fully described on page **404**.

The following species produce giam; those underlined usually dominate commercial supplies.

*Hopea helferi, H. nutans, H. pentanervia, H. semicuneata, H. apiculata, H. pierrei, H. polyalthioides, H. resinosa* and *H. subalata.*

### The tree
A large tree some 40m tall, with a diameter of about 1.2m.

### The timber
The sapwood is not very distinct from the heartwood unless discoloured by sap-stain. The colour of the heartwood is yellowish-brown, sometimes with a greenish tinge, darkening on exposure to a dark tan-brown. Planed surfaces are not particularly lustrous, and there is an occasional stripe figure on quarter-sawn surfaces. The grain is spiral, wavy or interlocked, and the texture fine and even. The timber is hard and heavy, weighing from 880 kg/m³ to 1040 kg/m³ when dried.

### Drying
Dries slowly and requires care if surface checking is to be avoided. Has a medium shrinkage rate.

### Strength
Similar to kapur (*Dryobalanops* spp.), but rather less tough.

### Durability
Very durable.

### Working qualities
Rather difficult to work, and tends to dull cutting edges fairly quickly, due mainly to its hard and horny nature. The tendency to spiral grain in some pieces, and the general interlocked grain may lead to tearing out during planing and moulding, but the wood is capable of a good finish, and it can be stained and polished resonably well.

364

## Uses

Giam is used for structural work, sleepers, boat building, flooring, textile rollers, and to some extent for furniture in the countries of origin.

# JELUTONG

*Dyera costulata* Hook f.                    Family: Apocynaceae

## Other names

jelutong bukit, jelutong paya (Sarawak).

## Distribution

Widely distributed in Malaysia, but nowhere abundant, it occurs in forests from sea level to 420m. It is also found in Sumatra and East Kalimantan.

## The tree

The trees grow to a very large size, up to 60m in height, and a diameter of up of 2.4m with straight and cylindrical boles often 27m long.

## The timber

The sapwood and heartwood are not differentiated by colour. The wood is creamy-white when first cut turning a pale straw colour after exposure. The planed wood is lustrous or very lustrous, without any figure, but slit-like radial passages may be seen on tangential surfaces as lens-shaped bodies. These are latex traces, and are commonly found in rows or clusters, often about 1.0m apart. The latex passages are commonly from 3mm to about 5mm wide, and about 12mm in length, and represent a natural defect which makes it difficult to use the timber in long lengths and wide widths where appearance is a primary consideration. They are, however, easily eliminated during conversion into smaller dimensions. The grain is usually straight, and the texture is fine and even. The wood is moderately soft, and moderately light, weighing about 470 kg/m$^3$ when dried.

## Drying

Relatively easy to dry, but difficulty may be experienced in extracting moisture from the centre of thick stock before staining occurs. There is little tendency for the wood to warp or split.

## Strength
A soft, weak and rather brittle timber.

## Durability
Non-durable.

## Working qualities
An easy timber to work with both hand and machine tools. It can be finished to a very smooth surface, takes stains, polish and varnish very well, and nails, screws and glues satisfactorily.

## Uses
Pattern-making, drawing boards, carving, wooden clogs, and for battery separators provided the timber is selected free from latex traces. It should be noted that jelutong possesses chemical properties that are deleterious to lead accumulators. This can be overcome by soaking the separators for 24 hours in a 5 per cent solution of caustic soda followed by a thorough washing in running water. Jelutong has other possible uses as core stock for flush doors, plywood, and for lightweight partitions and some forms of interior joinery.

# KAPUR

*Dryobalanops* spp.                    Family : Dipterocarpaceae

Various species of the genus *Dryobalanops* are distributed over part of Malaya, Sumatra, and Borneo, including Sarawak, Brunei, Sabah and East Kalimantan.
The various species comprising kapur are given below.

*Dryobalanops aromatica* Gaertn. f. and *D. oblongifolia* Dyer., produce Malaysian kapur for export, but locally the latter species is known as keladan.
*Dryobalanops aromatica* Gaertn. f., *D. lanceolata* Burck., and possibly other species produce Sarawak kapur.

366

*Dryobalanops lanceolata* Burck., and *D. beccarii* Dyer., and possibly other species produce Sabah kapur.

*Dryobalanops aromatica* Gaertn. f. *and D. oblongifolia* Dyer., are the principal species producing Indonesian kapur shipped from Sumatra.

*Dryobalanops aromatica* Gaertn. f., *D. oblongifolia* Dyer., *D. beccarii* Dyer., *D. fusca* V.Sl. and *D. lanceolata* Burck, and possibly other species produce Indonesian kapur shipped from Borneo.

## The tree
The trees vary according to species and environment, but usually they are very large, often reaching a height of 60m and a diameter of 1.5m to 2.0m with slightly tapering boles some 30m long above the buttresses. Although tending to grow gregariously, it never forms pure stands.

## The timber
The sapwood is whitish to yellowish-brown in colour, up to 75mm wide, and clearly defined from the heartwood, which is a light rose-red when freshly cut, becoming rose-red to reddish-brown on exposure, often (particularly when fresh) with a pronounced camphor-like odour. It is fairly straight grained, moderately coarse but even textured, and is moderately hard and heavy.

There is often a superficial resemblance to keruing, but on examination of end grain it will be seen that kapur has continuous gum ducts in concentric lines. These individual canals are smaller than the vessels, and are often filled with white resin; this does not exude over the surface of the wood.

The different species vary somewhat in weight, but on average they are as follows,

| | |
|---|---|
| Malaysian kapur | 790 kg/m$^3$ |
| Sarawak kapur | 750 kg/m$^3$ |
| Sabah kapur | 750 kg/m$^3$ |
| Indonesian kapur | 590 to 830 kg/m$^3$ |

all when dried.

## Drying
Reported to dry fairly well with only a slight tendency to warp and check.

## Strength
From the results of limited tests it would seem that in the green state kapur is only slightly superior to teak in general strength properties, but when dried, it is harder, and some 15 per cent stronger in bending, and about 50 per cent stiffer and more resistant to suddenly applied loads than teak.

## Durability
Very durable, according to UK test reports, but classified as only moderately durable in the countries of origin.

## Working qualities
Works moderately well with both hand and machine tools, but is variable in respect of dulling of cutting edges, but on the whole kapur is slightly better than the average run of keruing. Cutters should be kept well sharpened to avoid trouble from raised grain and to obtain a smooth finish. Takes nails, screws, stains and polish satisfactorily.

## Uses
Constructional purposes, and domestic or light-duty flooring. The wood is liable to stain if allowed to come in contact with iron in moist conditions.

# KATON

*Sandoricum indicum* Cav.                     Family: Meliaceae

## Other names
sentul (Malaysia); kalampu (Sarawak and Brunei); thitto (Burma).

## Distribution
Occurs in the evergreen forests of Thailand, Malaysia, Sarawak, Brunei, Indonesia and Lower Burma.

## The tree
A medium to large-size tree, attaining a height of 36m in favourable conditions and a diameter of about 0.75m with straight, clear boles of 18m to 21m above the fluted base.

368

## The timber
The sapwood is pale pink, and not always clearly defined from the heartwood which is pink in colour when freshly cut, darkening to a light reddish-brown on exposure. Occasionally broken stripe or fiddleback figure is present, otherwise the grain is straight to slightly interlocked. The texture is fine and even, and the wood is moderately hard and moderately heavy, and weighs about 580 kg/m³ when dried.

## Drying
Requires care in air drying to avoid surface checking, and in kiln drying to avoid collapse in thick material.

## Strength
No information is available.

## Durability
Moderately durable.

## Working qualities
Reported to be easy to work and saw, and takes a fine, smooth finish. Quarter-sawn material may tend to pick up during planing and moulding when interlocked grain is present. Nails, screws and glues reasonably well, and takes stains and polish satisfactorily.

## Uses
Joinery, furniture, light interior construction, boat building.


# KELAT

*Eugenia* spp.                                  Family: Myrtaceae

About 20 species of the genus *Eugenia* are used in Malaysia in the production of the medium hardwood kelat. The principal species found in commercial supplies are *E. cerina*, *E. flosculifera*, *E. glauca*, *E. griffithii*, and *E. longiflora*.

## General characteristics
The heartwood is greyish-brown, deep brown, or reddish-brown, and the sapwood is usually of a lighter colour but not

always well defined. The wood is without lustre or figure, the grain is interlocked, irregular, or wavy, and the texture throughout the range is generally moderately fine. The weight varies considerably with the species but the endeavour is to use those where the density is over 640 kg/m³ but less than 960 kg/m³ when dried.

The wood is fairly easy to work, but there is usually a tendency for the grain to pick up in planing and moulding. The wood is capable of a smooth finish, and it takes stains and polish quite well.

## Uses
Kelat is moderately durable, and is suitable for general construction and joinery.

# KELEDANG

*Artocarpus* spp.                                    Family: Moraceae

Various species of the genus *Artocarpus* occur throughout Eastern and South East Asia, certain of those in the latter area produce Malaysian keledang, principally *Artocarpus lanceifolius* Roxb., but any of the following may be included.

*A. anisophyllus, A. dadah, A. fulvicortex, A. heterophyllus A. incisus, A. integer, A. kemando, A. maingayii, A. nitidus* and *A. rigidus.*

## General characteristics
The sapwood is moderately to well defined from the heartwood which is orange-yellow-brown, turning on exposure to gold-brown, dark orange-brown, walnut-brown, or russet-brown according to the species. Planed surfaces are moderately lustrous or very glossy, and there is a stripe figure on radial surfaces. The grain is usually deeply interlocked, and the texture moderately coarse but even. Chalky white deposits are often present in the vessel cavities. The wood varies in weight from 560 kg/m³ to 800 kg/m³ or more, when dried.

## Drying
No information is available, but it is reputed to dry without difficulty, and this is perhaps substantiated by the end-uses to which keledang is applied.

## Strength
Although variable, because of the different densities, on average the strength properties are similar to those of kapur (*Dryobalanops* spp.).

## Durability
Moderately durable to durable.

## Working qualities
Works and finishes fairly easily, although the interlocked grain may give trouble in planing and moulding. Takes a good polish.

## Uses
Bridge construction, cooperage, coffin boards, furniture and for louvre shutters.

# KEMPAS

*Koompassia malaccensis* Maing        Family: Leguminosae

## Other names
Kempas appears to be the commercial description for the wood of *K. malaccensis*. *K. excelsa* is generally described separately eg tualang (Malaysia), kayu raja (Sarawak) and mengaris (Borneo generally).

## Distribution
The genus *Koompassia* has a fairly general distribution in Malaysia, Sumatra and Borneo (East Kalimantan). It is found in the low hill forests and swampy areas.

## The tree
A large tree with large buttresses, reaching a height of about 54m and a diameter of 1.2m although larger specimens are found occasionally. Clear boles 24m to the first branch are common.

## The timber
The sapwood is well defined, white to pale yellow in colour, and about 50mm wide in large trees. The heartwood is brick-red when freshly sawn, weathering to orange-red speckled by

yellow-brown lines or streaks due to the soft parenchymatous tissue which surrounds the pores. The wood is moderately lustrous, the grain is typically interlocked, sometimes spiral or wavy, and the texture is coarse and even. The weight is about 880 kg/m³ when dried.

Concentric rings of abnormal tissue (included phloem) may be visible on the cross-section of unsawn logs. These are very hard and horny to cut across the grain, but to a great extent this is discarded during sawing into lumber. However, occasional streaks may be observed in converted stock; they appear as veins of hard, stone-like tissue, usually about 6mm wide, and may extend for one metre or more in the direction of the grain. They represent a source of mechanical weakness which can lead to splitting in drying, and they limit the strength of the wood.

## Drying
Dries reasonably well, but with some tendency to warp and check. Where hard zones of abnormal tissue are present the tendency to split is much greater.

## Strength
Similar to karri (*Eucalyptus diversicolor*) in most strength categories but with greater resistance to crushing loads.

## Durability
Durable.

## Working qualities
Rather difficult to work because of its density and hardness, the blunting effect on tool edges often being severe. Straight-grained material can be planed and moulded to a good finish, but quarter-sawn stock generally tends to tear unless a cutting angle no greater than 20° is employed. Kempas takes a good finish but may require filling before polishing. Takes nails and screws fairly well but pre-boring assists this operation.

## Uses
Heavy constructional work, bridge building, sleepers (treated), core stock for plywood, and shingles from selected straight grained material. The wood has a slightly acidic character and may encourage the corrosion of ferrous metals. These should be protected from contact with the wood.

The closely related tualang (*Koompassia excelsa* Taub.) has a greyish-white sapwood, distinct from the heartwood which is reddish-brown or dark red-brown when freshly cut, turning dark brown on exposure. The wood has a stripe figure on radial surfaces, and a feathery tracery on tangential surfaces caused by wavy confluent bands of parenchyma (usually much more pronounced than in *K. malaccensis*). The phenomenon of included phloem occurs in tualang as in kempas. The grain is interlocked, and the texture coarse but even. The wood weighs about 800 to 864 kg/m$^3$ (rather less than kempas), and is probably moderately durable. Tualang has similar uses to kempas.

## KERANJI

*Dialium* spp.                                    Family: Leguminosae

Several species of *Dialium* are used in Malaysia in the production of commercial keranji, those normally dominating supplies are underlined in the list given here.
*Dialium kingii, D. kunstleri, D. laurinum, D. maingayi, D. patens, D. platysepalum,* and *D. wallichii.*

### General characteristics
The sapwood is light brown in colour, distinct from the heartwood, and may be 75mm or more wide. The heartwood varies according to species, from golden-brown to red-brown or dark red-brown, weathering to darker shades. Planed surfaces are generally very lustrous, and there is usually a stripe figure on quarter-sawn surfaces, and light-coloured zig-zag markings on tangential surfaces caused by abundant concentric bands of parenchyma. The grain is wavy and interlocked, and the texture moderately fine to moderately coarse and even. The wood is hard and heavy, the weight ranging from 800 to 1120 kg/m$^3$ when dried.
Keranji is difficult to work, due to its hard and horny nature, and is moderately durable in severe conditions of external exposure.

### Uses
A hard, heavy, strong wood, generally suitable for structural purposes and underwater construction.

# KERUING

*Dipterocarpus* spp.                    Family : Dipterocarpaceae

There are some thirty species of *Dipterocarpus* occurring in South East Asia. For practical purposes it has been considered best to give the description and properties of the timbers together, since the exporting countries do not differentiate between the species but sell them collectively. This may result in variations in colour, weight etc. Some of the more important commercial species are given below.

*Dipterocarpus cornutus* Dyer, *D. costulatus* V..Sl, *D. crinitus* Dyer, *D. sublamellatus* Foxw. There may also be included, *D. apterus, D. lowii, D. verrucosus,* and others.
These produce Malaysian keruing.

*Dipterocarpus cornutus* Dyer, and *D. gracilis* Bl. produce Indonesian keruing.

*Dipterocarpus acutangulus* Vesque, *D. apterus* Foxw., *D. caudiferus* Merr, *D. lowii* Hook f., and *D. verrucosus* Foxw ex V.Sl.
These produce Sarawak keruing.

*Dipterocarpus acutangulus* Vesque, *D. caudiferus, D. confertus* V.Sl., *D. gracilis* Bl., *D. grandiflorus*, Blanco, and *D. warburgii* Brandis.
These produce Sabah keruing.

## Other names
Malaysian, Indonesian, Sarawak, Sabah gurjun, kruen.

## Distribution
The trees of this genus occur gregariously throughout South East Asia, Burma, India, The Andamans, Sri Lanka, Thailand, South Vietnam, Cambodia and the Philippines.

## The tree
The trees are evergreen and generally large, depending on locality, reaching heights of 25m to 45m and a diameter of 0.75m to 1.5m with a straight, cylindrical bole often clear of branches to 21m above a small buttressed base.

374

## The timber

The sapwood is greyish-brown and distinct, and usually 50mm to 75mm wide. The heartwood varies from light red to reddish-brown or brown, darkening on exposure. The grain is usually straight but may be shallowly interlocked, and the texture is moderately coarse or coarse, but even. The timber contains an oleo-resin which exudes in a number of pieces, especially on the end grain. The various species vary in weight from 640 kg/m³ to 960 kg/m³ but an average for the group is about 740 kg/m³ when dried.

## Drying

The timbers of this group dry slowly, and it is difficult to achieve uniform drying particularly in thick stock or in quarter-sawn material. High temperatures aggravate the exudation of resin, and care is therefore needed in order to reduce this to a minimum and also to reduce the tendency for cupping to occur. There is high shrinkage during drying, and a corresponding high movement potential.

## Strength

Keruing generally compares with teak in most strength categories, but it is a little stiffer, and about 40 per cent more resistant to shock loads.

## Durability

Moderately durable.

## Working qualities

The variations that exist between the different species and within species in respect of resin content, type of grain and hardness, affect the general ease of working, dulling of cutters, and finishing properties. The timbers vary from hard to very hard and horny in cutting across the grain, and whereas straight grained material is capable of clean, slightly fibrous finish in planing and moulding, quarter-sawn, or interlocked-grained material picks up during these operations unless cutting angles are reduced to 20°. The wood stains satisfactorily, but due to the resin, varnishing and polishing require a lot of care. It takes nails and screws reasonably well, but its gluing properties are variable.

## Uses
Heavy and light construction, wharf decking, bridges, flooring, sills, wagon sides and floors, vehicle building. For durable exterior construction, the timber should be treated with preservative.

# KUNGKUR

*Pithecellobium confertum,*                    Family: Leguminosae
*P. splendens*, and *P. bubalinum*

Various species of *Pithecellobium* occur throughout South East Asia, those in Malaysia producing a light hardwood known as kungkur.

## General characteristics
The sapwood is yellowish-brown, and is sharply demarcated from the heartwood which varies according to species from deep pink to light red or dark reddish-brown. Planed surfaces are lustrous or very lustrous, without figure due to the generally featureless character of the wood, and the lack of contrast between the very fine or moderately fine rays and the background. The grain is wavy or interlocked, and the texture coarse and rather uneven. Kungkur weighs about 672 kg/m$^3$ when dried.
Moderately durable, the wood is easy to saw, but rather troublesome in planing and moulding due to the interlocked grain, but it is capable of a good finish.

## Uses
Joinery and light construction.

# MACHANG

*Mangifera* spp.                    Family: Anacardiaceae
principally *M. foetida* Lour.

Machang from Malaysia is the product of several species of *Mangifera*, and although *M. foetida* usually predominates commercial shipments, the following may be included, in one or more species.

*M. indica, M. lagenifera, M. longipes, M. longipatiolate, M. microphylla,* and *M. quadrifida.*

## General characteristics
There is little difference in colour between sapwood and heart-wood, the wood generally being light brown, light greyish-brown or pinkish-brown in colour, occasionally with a core of chocolate-brown or black streaked with yellow. Planed surfaces are lustrous or very lustrous, and radial surfaces often display a stripe figure, while there are zig-zag markings on tangential surfaces, produced by banded parenchyma. The grain is inter-locked and wavy, and the texture moderately coarse or coarse and even. The wood weighs about 560 kg/m$^3$ when dried.
Machang is non-durable, and is easy to saw, but rather difficult to obtain a smooth finish in planing and moulding. With sanding, it is capable of a good finish and takes a good polish but requires a reasonable amount of filling.

## Uses
Interior joinery and light construction.

# MEDANG

### Species of Lauraceae

A large number of species of the Lauraceae family produce light-weight to medium-weight timbers with similar character-istics and properties. The following species are used in Malaysia to produce medang, those underlined are the principal species, but one or more of the other species may also be included in shipments of the timber.

| | |
|---|---|
| *Actinodaphne sphaerocarpa* | *C. javanicum* |
| *Alseodaphne coriaceae* | *C. porrectum* |
| *A. insignis* | *C. velutinum* |
| *A. pendulifolia* | *Cryptocarya bracteolata* |
| *A. peduncularis* | *C. griffithiana* |
| *Beilschmiedia insignis* | *C. kurzii* |
| *B. praecox* | *C. tomentosa* |
| *B. tonkinensis* | *Dehaasia nigrescens* |
| *Cinnamomum cinereum* | *D. cuneata* |
| *C. iners* | *D. curtisii* |

377

D. elliptica
Dehaasia incrassata
Litsea castanea
L.curtisii
L. finestrata
L. ferruginea
L. firma
L. gracilipes
L. grandis
L. machilifolia
L. megacarpa

L. maingayi
L. medularis
L. petiolata
L. robusta
L. tomentosa
Nothaphoebe panduriformis
N. umbelliflora
Phoebe declinata
P. grandis
P. macrophylla

## The timber
With so many possible species being included there is a wide variation in colour. The sapwood is usually fairly to moderately well-defined from the heartwood, which may be light-straw, yellow-brown, pink-brown, red-brown, olive-brown, or olive green, with the majority having an olive tinge. The planed surfaces are generally dull, most species are without figure, but a few display an attractive zig-zag figure on tangential surfaces due to banded parenchyma. Some have a distinct aromatic odour.
The grain on the whole is interlocked or wavy, and the texture is mostly moderately fine and even, but in some cases moderately coarse. The weight of the various species ranges from 400 to 800 kg/m³ but the majority are in the range of 560 to 640 kg/m³ when dried.

## Drying
Medang on the whole is reputed to dry without undue degrade and with very low shrinkage.

## Strength
Australian sources give strength properties of *Cinnamomum* spp., as roughly comparable with those of white meranti, but about 25 per cent harder, and very much tougher than that timber.

## Durability
While a few of the species have a good reputation for durability, generally medang should be considered only moderately durable.

## Working qualities
Medang works easily and well, although there is some variation between the species; some are soft, others are moderately hard to hard, but there are no deposits in the wood, and the interlocked or wavy grain appears to be the main factor in planing and moulding. The wood takes a good polish and nails and screws well.

## Uses
Joinery, cabinet-making, interior construction, panelling, lining and plywood.

# MELUNAK

*Pentace triptera* Masters                    Family: Tiliaceae

Although *Pentace triptera* dominates commercial shipments of melunak from Malaysia, two further species may be included, ie, *P. curtisii* and *P. macrophylla*.

## General characteristics
The sapwood is lighter coloured, but not sharply defined from the heartwood which varies from pinkish-brown to light red-brown or red-brown in colour. Planed surfaces are without lustre, and there is a stripe figure on radial surfaces. The grain is interlocked, and the texture moderately fine and even. White deposits are occasionally present in the cells, and the wood displays distinct ripple marks. Melunak weighs about 656 kg/m$^3$ when dried.

## Drying
Reported to dry slowly, with little degrade.

## Strength
Compares favourably with American mahogany (*Swietenia*), but is harder.

## Durability
Moderately durable.

## Working qualities
Moderately easy to work, but the interlocked grain is liable to pick up when planing or moulding quarter-sawn stock. A reduction of the cutting angle to 20° is usually beneficial. It takes nails, screws and polish very well, but the rather large pores generally require a lot of filling.

## Uses
Melunak closely resembles the related thitka (*P. burmanica*) of Burma, and is used for similar purposes, but the larger pores in melunak affect the finishing properties to a greater extent than with thitka. In South East Asia, melunak is used for furniture, flooring, panelling, boat building and for shop fittings.

# MEMPISANG

## Species of Anonaceae

A number of related species of the Anonaceae family occur in Malaysia and Indonesia. Those generally employed to make up mempisang from Malaysia are as follows, the dominating species being underlined.

*Mezzettia leptopoda* Oliv.　　*Polyalthia glauca*
*Monocarpia marginalis*　　　*Platymitra siamensis*
*Cathocalyx carinatum*　　　　*P. hypoleuca*
*C. pahangensis*　　　　　　　*Xylopia ferruginea*
*X. fusca.*

## General characteristics
The sapwood is not differentiated from the heartwood by colour; the heartwood is light yellow with a greenish or pinkish tinge. All the species are characterized by two distinct types of ray, the broader of which produces a distinctive silver ray figure on radial surfaces. The grain is fairly straight throughout the range, and the texture generally coarse and uneven. The weight is variable, from 480 to 800 kg/m$^3$ but commercial parcels usually average about 672 kg/m$^3$ when dried.
Mempisang is moderately difficult to dry, and care is needed to prevent surface checking and end splitting. It is also rather

380

difficult to work, but much depends upon the actual species. It is not unlike European oak in its general working and finishing properties, although some species are much softer than oak.

## Uses
Mempisang is not durable, but is suitable for interior construction, joinery and furniture.

# MENGKULANG

*Heritiera* spp.                                    Family: Sterculiaceae
syn. *Tarrietia* spp.

## Other names
chumprak, chumprag (Thailand) ; kembang (Sabah).

## Distribution
Various species of the genus *Heritiera* occur throughout South East Asia, and of the six that are found in Malaysia, three species are more commonly used in the production of commercial mengkulang, ie, *H. simplicifolia, H. javanica,* and *H. borneensis,* the first named usually dominating the selection.

## The tree
A fairly large tree, attaining a height of 36m to 42m and a diameter of about 0.6m although some specimens may reach 1.0m or slightly more. The boles are generally straight and cylindrical and clear for about 24m above the large buttresses.

## The timber
The sapwood is yellow to light brown in colour, clearly defined from the heartwood, which is pinkish-brown, but more commonly reddish-brown or dark brown. Planed surfaces are lustrous, and quarter-sawn surfaces are marked with moderately conspicuous reddish flecks caused by the fairly large rays. The grain is straight to shallowly interlocked, and the texture is coarse and fairly even. The wood has a slightly greasy feel, and when fresh, has a slightly unpleasant odour. Ripple marks are commonly present. The wood is moderately hard, and moderately heavy, weighing about 720 kg/m$^3$ when dried.

## Drying
Dries fairly easily, but with a tendency to warp and twist.

## Strength
Similar to teak in all strength categories.

## Durability
Moderately durable.

## Working qualities
The wood works with moderate ease, but tends to blunt tools, especially saw teeth, fairly quickly. The timber can be planed to a smooth finish but care is needed when planing quarter-sawn material in order to avoid the grain picking up. Mengkulang peels well, and glues satisfactorily, but the wood is liable to split when nailed. It is said that the material from Malaysia is slightly more difficult to work than that from Thailand. Mengkulang takes paint and polishes well.

## Uses
General construction, joinery, plywood, panelling, sills, sleepers, and for paper-pulp. It is also suitable for flooring of the domestic type; under heavy use, the surface tends to break down with a splintery effect.


# MERANTI

*Shorea* spp.                              Family: Dipterocarpaceae


The genus *Shorea* produces a number of distinct commercial timbers, some of which are provided by several botanical species. The nomenclature is somewhat involved as the trade names vary according to the country of origin. The timbers are classified for export mainly on the basis of colour and density.
It should also be noted that the botanically associated genera, *Parashorea* and *Pentacme* produce similar timber to meranti, and may be included with *Shorea* in commercial parcels under a separate trade name. For example,
**Meranti** is the group name for species of *Shorea* in Malaysia, Sarawak, Brunei and Indonesia.

**Seraya** is the group name for species of *Shorea* and *Parashorea* in Sabah.

**Lauan** is the group name for species of *Shorea*, *Parashorea* and *Pentacme* in the Philippines.

These are all light to medium-weight timbers.

The heavy species of *Shorea* belong to the **balau** group of Malaysian species, and this group is roughly equivalent to **selangan batu**, which is the name used for similar timbers in Sarawak and Sabah. Other timbers in this group are **chan** of Thailand and **sal** of India.

There is considerable variation in colour of the wood in certain species, and for this reason may be classed as light or dark according to the colour of a particular piece. Thus the lighter-coloured timber of *Shorea acuminata* in Malaysia is commonly classed **light red meranti** while the darker-coloured wood is included with **dark red meranti**.

The following notes describe the timbers of the various groups, together with their botanical species. Wherever possible, those species that dominate commercial supply are underlined.

## Light red meranti

*Shorea acuminata* Dyer (in part)
*S. leprosula* Miq.
*S. parvifolia* Dyer
*S. macroptera* Dyer
*S. ovalis* Bl
*S. dasyphylla*
*S. lepidota*

*S. palembanica* Miq.
*S. teysmanniana* Dyer ex Brandis
*S. platycarpa* (in part)
The above species produce light red meranti from Malayasia.

*Shorea albida* Sym.
*S. parvifolia* Dyer.
*S. quadrinervis* V.Sl.
are the principal species producing light red meranti from Sarawak and Brunei. The timber from these species is also known in Sarawak as perawan, and, more specifically, the lighter-weight specimens of *S. albida* are known as meranti bunga or alan bunga.

*Shorea leprosula* Miq.
*S. ovalis Bl.*
*S. parvifolia* Dyer.
are the principal species producing light red meranti from Indonesia, also locally known as red meranti and lauan.

*Shorea leprosula* Miq.
*S. leptoclados* Sym.
*S. parvifolia* Dyer.
*S. smithiana* Sym.
are the principal species producing light red seraya from Sabah,
also locally known as red seraya and seraya merah.

**Dark red meranti**
*Shorea pauciflora* King.
*S. curtisii* Dyer ex King (in part)
*S. acuminata* (in part)
*S. platycarpa* (in part)
*S. platyclados*

are the principal species producing dark red meranti from Malaysia. The alternative Malaysian name for the wood of *S. pauciflora* is **nemesu**.
*Shorea pauciflora* King. is the principle species producing dark
red seraya from Sabah, locally known also as **oba suluk**.

**Yellow meranti**
*Shorea faguetiana* Heim
*S. resina-nigra* Foxw.
*S. balanocarpoides*
*S. gibbosa* Brandis
*S. hopeifolia* (Heim) Sym.
*S. maxima* Sym.
*S. multiflora* Sym.
These species produce the yellow meranti of Malaysia, or
meranti damar hitam.

*Shorea faguetiana* Heim.
*S. hopeifolia* Sym.
*S. multiflora* Sym.
These species produce the yellow meranti of Brunei, and
Sarawak, where the local name is **lun**, or **lun kuning**.

384

*Shorea acuminatissima* Sym.
*S. faguetiana* Heim.
*S. gibbosa* Brandis
These are the principal species producing yellow seraya of Sabah, where the local names are also **seraya kacha**, and **selangan kacha**.

## White meranti
*Shorea bracteolata* Dyer.
*S. hypochra* Dyer.
*S. assamica* Dyer.
*S. bentongensis.* Foxw.
*S. dealbata.*
*S. lamellata.*
*S. resinosa.*
*S. sericeiflora.*
*S. talura.*
These are the principal species producing white meranti from Malaysia, Sarawak, Brunei and Sabah, also known as **lun** or **lun puteh** in Sarawak and **melapi** in Sabah.
It should be noted that white meranti is not the equivalent of Sabah white seraya (*Parashorea* spp.).

### The tree
The various species of *Shorea* grow to a height of 45m or so and a diameter of 1.2m or a little more, with long, clean, cylindrical boles above small buttresses.

### The timber
The various types of commercial meranti/seraya show considerable variation in colour, density, and general properties, according to the species, and the following descriptions must therefore be regarded as being average for each group.

### Light red meranti/seraya
Sapwood lighter-coloured and usually distinct from the heartwood which is light-red or pinkish-brown; planed surfaces fairly lustrous, stripe figure on radial surfaces, with subtle, but attractive speckles caused by the rays. Texture rather coarse, but even, grain interlocked and wavy. Resin canals which may, or may not be plugged with white resin occur in tangential lines on end grain surfaces, but the wood is not resinous like keruing.

The wood varies in weight according to species from 400 kg/m$^3$ to 705 kg/m$^3$ but averages about 550 kg/m$^3$ when dried.

## Durability
Moderately durable.

## Dark red meranti/seraya
Sapwood lighter in colour and distinct from the heartwood which is red-brown darkening to a dark red; planed surfaces fairly lustrous, stripe figure on radial surfaces. Grey-coloured narrow streaks are often present on all longitudinal surfaces, caused by concentric layers of resin canals. The texture is rather coarse but even, and the grain is interlocked and wavy. The wood weighs on average, 710 kg/m$^3$ when dried.

## Durability
Durable.

## Yellow meranti
Sapwood lighter in colour and moderately distinct from the heartwood, which is light-yellow-brown, often with a greenish or olive tinge, weathering to a light brown colour; planed surfaces without lustre, faint stripe figure on radial surfaces. The texture is moderately fine or moderately coarse but even, and the grain is interlocked. It varies in weight from 576 kg/m$^3$ to 736 kg/m$^3$ but averages about 660 kg/m$^3$ when dried.

## Durability
Moderately durable.

## White meranti
Sapwood ill-defined or indistinct when freshly cut but becoming well defined in the course of drying; heartwood almost pure white when freshly cut, gradually changing to yellow-brown or buff colour, and weathering to a golden-brown or definite brown; planed surfaces lustrous with a subtle ribbon figure on radial surfaces. The texture is moderately coarse but even, and the grain is interlocked. The rays are inconspicuous on the radial surface. Vertical canals of the concentric type filled with white resin appear on end grain surfaces, but since they are smaller in size than the vessels they are not conspicuous on longitudinal surfaces. The wood weighs on average 660 kg/m$^3$ when dried.

## Durability
Moderately durable.

## Drying
The various types of meranti/seraya are reported to dry rapidly and well, with little degrade. Some slight distortion and surface checking may occur in the denser types. 'Malayan Forest Service Trade Leaflet No 8' gives the following information regarding the air drying times for red meranti dried under cover in Malaysia. From about 60 per cent moisture content to 18 per cent moisture content:—

| | |
|---|---|
| 25mm boards | 2 to 3 months |
| 38mm boards | $3\frac{1}{2}$ to $4\frac{1}{2}$ months |
| 50mm boards | approximately 5 months. |

## Strength
There is a wide variation in the strength properties of the various merantis and red seraya due to the differences in density and the number of species involved. Large, over-mature logs are frequently spongy in the heart, the wood in these areas being weak and brittle. Despite the fact that the best type of light red meranti is almost equal in strength to the weakest type of dark red meranti, there is nevertheless on average, a distinct difference in mechanical properties. The average figure for strength and stiffness in bending and compression for dark red meranti is about 20 per cent higher than that for light red meranti; in shear there is about 10 per cent difference, and in hardness, over 30 per cent. The Malayan Forest Service prepared the following table, on the basis of mechanical test results, where the mechanical properties of six other timbers are compared with those for light red meranti, the data for which are expressed in each case as 100.

| Timber | Maximum load in bending | Modulus of elasticity | Maximum crushing strength | Side hardness | End hardness | Shear |
|---|---|---|---|---|---|---|
| light red meranti | 100 | 100 | 100 | 100 | 100 | 100 |
| dark red meranti | 122 | 121 | 125 | 139 | 131 | 111 |
| Central American mahogany | 118 | 84 | 112 | 116 | 112 | — |
| sapele | — | 96 | 120 | 169 | 157 | — |
| Scots pine | 76 | 86 | 74 | 77 | 69 | 87 |
| oak | 105 | 86 | 101 | 214 | 181 | 135 |
| teak | 146 | 108 | 145 | 186 | 137 | 122 |

According to these values, light red meranti is almost equal to oak in strength properties, but oak is much harder, while Scots pine has only about 75 per cent of the general strength of light red meranti.

White and yellow meranti are reported to have similar strength properties to those of American mahogany, but with a slightly lower resistance to splitting in the tangential plane in the case of white meranti.

## Working qualities

The wood of the various species work well and in general are capable of a good smooth surface, but a reduction of cutting angle to 20° is beneficial where a tendency for the grain to tear becomes apparent. The dulling effect on saws and cutters varies somewhat with the species, but is usually quite small, except in the case of white meranti which generally contains a fairly high amount of silica in the ray cells. The various species can be glued, nailed and screwed satisfactorily, and can be stained and polished quite well after suitable filling.

## Uses

All the merantis are well suited to joinery and general construction, and also for furniture, particularly for interior framing and drawer sides and backs, but more specifically, the end uses are as follows,

**Light red meranti/light red seraya.** Joinery, plywood, flooring, packing cases, panelling, carpentry, cheaper grade furniture, and general construction.

**Dark red meranti.** As above, but higher durability factor allows for its use in more exposed situations, for example for cladding, particularly when treated with a preservative.

**Yellow meranti.** Joinery, light construction, interior finish, plywood, furniture.

**White meranti.** Flooring, railway carriage framing, furniture, shop fitting, joinery, ship and boat planking, veneer.

See also **meranti gerutu, white seraya** and **selangan batu.**

# MERANTI GERUTU

*Parashorea* spp.            Family : Dipterocarpaceae
principally *P. lucida* Kurz.

## Other names
gerutu gerutu or gerutu.

Three species of *Parashorea* producing meranti gerutu occur in South East Asia, but because of the restricted distribution of two of them, the bulk of sawn wood that is shipped is that of *P. lucida*. The name meranti gerutu has been adopted in Malaysia for the wood, but it is shipped from West Malaysia and Singapore either under that name, or by either of the alternatives given above.

## General description
The timber possesses affinities with the red meranti group of *Shorea* spp., and resembles light red meranti in colour, and dark (heavy) red meranti in strength properties.
The heartwood is very light brown or light bronze with a distinct pink tinge, darkening on exposure. The grain is deeply interlocked making the timber difficult to split radially, and the texture is rather coarse but even. The wood weighs from 672 kg/m³ to 768 kg/m³ and averages 690 kg/m³ when dried.

## Drying
Dries easily and well with only a slight tendency to warp and check.

## Strength
Refer to dark red meranti. The strength of meranti gerutu is similar.

## Durability
Moderately durable.

## Working qualities
The wood works quite well although there is some resistance in sawing, generally of the same order as for kapur and keruing. Planer knives and cutters must be kept sharpened if a good finish is to be obtained. Takes nails and screws well, and stains and polishes reasonably well.

## Uses

Interior construction, joinery, vehicle flooring, light industrial flooring. A similar species of *Parashorea* is used in Burma for boat building.

# MERAWAN

*Hopea* spp.                                    Family: Dipterocarpaceae

## Other names

selangan (Sarawak and Sabah).

**Merawan** is the group trade name used in Malayasia for several species of the genus *Hopea*, similarly, the group name in Sarawak and Sabah for these species is **selangan**. This latter trade name should not be confused with the current classification of **selangan batu** (described herein under that name), nor with **giam**, the Malaysian and Sarawak trade name for a further group of *Hopea* spp., also known in Sabah as **selangan batu**.

The following species generally produce merawan from Malaysia, those underlined normally dominating commercial supplies.

*Hopea griffithii, H. mengarawan, H. nervosa, H. odorata, H. pubescens, H. sangal, H. sulcata, H. beccariana, H. dryobalanoides, H. dyeri, H. ferruginea, H. glaucescens, H. johorensis, H. latifolia, H. minima, H. montana, H. myrtifolia, H. pedicellata,* and *H. sublanceolata.*

## The tree

A large tree with a straight cylindrical bole, attaining a height of 40m or a little more, and a diameter of 1.2m above the buttresses. Clear boles of up to 24m are not uncommon.

## The timber

The sapwood is light yellow in colour and distinct from the heartwood which is yellow-brown in most species, olive-brown in some, and brown with a reddish tinge in a few when freshly cut, turning a dark brown or chocolate brown in course of time. The planed surfaces are mostly very lustrous; often with a stripe figure on radial surfaces and conspicuous white or grey-coloured narrow streaks on all longitudinal surfaces produced by vertical resin canals. The grain is interlocked, and the texture moderately fine and even. The weight varies with the species, ranging from 576 kg/m$^3$ to 848 kg/m$^3$, but commercial parcels generally average about 705 kg/m$^3$ when dried.

## Drying
Dries slowly, but well, with a tendency to surface check unless care is taken. Has a low to medium shrinkage rate.

## Strength
Although the strength properties vary with the species and their density, on the whole, the general strength properties of merawan are similar to those of kapur (*Dryobalanops* spp).

## Durability
Durable.

## Working qualities
Merawan is fairly difficult to saw and work; it is a hard to moderately hard timber, and with the interlocked grain, some dulling of cutters, and tearing of grain in planing and moulding is likely. Reduction of cutting angles to 20° is generally beneficial. The wood is however, capable of a smooth finish, and takes stain and polish satisfactorily.

## Uses
Structural uses, first-class joinery and furniture.

# MERBAU

*Intsia palembanica* Miq.                    Family: Leguminosae

## Other names
mirabow (Sabah).

## Distribution
Malaysia and Indonesia.

## The tree
A large tree, reaching a height of 42m and a diameter of 0.9m. The boles may be clear for about 18m above the large buttresses, but are not always straight.

## The timber

The sapwood, usually about 75mm wide, is sharply differentiated from the heartwood, and is whitish to pale yellow in colour. The heartwood is orange-brown or brown or dark red-brown, weathering to darker shades. Lighter-coloured parenchymatous markings often give the wood an ornamental figure on tangential surfaces. The wood is rather hard and heavy, weighing about 830 kg/m³ when dried. The texture is coarse but even, and the grain is interlocked and often wavy. Sulphur-yellow and dark-coloured deposits are characteristic of the species, and can commonly be seen in the vessel cavities. It is reported that the timber is liable to promote the corrosion of ferrous metals under moist conditions, but there is some doubt about this.

## Drying

The timber dries slowly without appreciable degrade, but if submitted to rapid drying conditions there is a definite tendency for end splitting and surface checking to occur.

## Strength

Merbau has good strength properties and is similar to mengkulang in this respect.

## Durability

Durable.

## Working qualities

The working properties of merbau are similar to those of West African afzelia, but the higher incidence of gum in merbau tends to collect on saws, and there is also a tendency for the grain to tear out in quarter-sawn material during planing and moulding. A reduction of cutting angle to 20° is beneficial. The wood tends to split in nailing, but holds screws well, and can be stained and polished reasonably well. The yellow deposits are soluble in water, the dye thus produced having a lasting effect on textiles.

## Uses

Merbau is used locally for heavy constructional work, sleepers, agricultural implements, axe and tool handles, and furniture. The wood has a moderate to high resistance to wear, and is therefore suitable for flooring, generally for low to medium concentrations of traffic. It has a relatively low shrinkage, but it does tend to split when nailed.

# MERPAU

*Swintonia* spp.                                    Family: Anacardiaceae

There are two species of *Swintonia* generally used in Malaysia to produce merpau, ie *S. schwenkii* and *S. spicifera*, but occasional specimens of *S. penangiana* may be included.

### General characteristics
The sapwood is not well defined but is a little lighter in colour from that of the heartwood, which varies from a grey-brown with or without a pinkish tinge, to reddish-brown. The grain is interlocked and produces a stripe figure on quarter-sawn surfaces, and wood parenchyma produces ornamental zig-zag markings on tangential surfaces. The texture is rather coarse but even.
Classified in Malaysia as a medium hardwood it weighs an average 752 kg/m³ when dried. It works moderately well and has good nailing properties, and is considered to be durable.

### Uses
General and marine construction above water, or in water where marine borer is not a problem.

## MERSAWA and KRABAK

*Anisoptera* spp.                                Family: Dipterocarpaceae

### Other names
The various species of *Anisoptera* are generally mixed and exported under the trade names of mersawa and krabak, according to the country of origin. The following species may be included in either group, but those which dominate commercial supplies are underlined.

Malaysia    *Anisoptera curtisii* Dyer  ⎫
            *A. laevis* Ridl.          ⎪
            *A. marginata* Korth.      ⎪
            *A. costata* Korth.        ⎬  mersawa
            *A. megistocarpa* V.Sl.    ⎪
            *A. oblonga* Dyer.         ⎪
            *A. scaphula* (Roxb.) Pierre ⎭

Thailand    *A. curtisii* Dyer.
            *A. oblonga* Dyer.
            *A. scaphula* (Roxb.) Pierre       } krabak
            *A. cochinchinensis*
            *A. glabra* Kurz

## The tree
The trees vary in size according to the species but generally attain a height of about 45m and diameters of 1.0m to 1.5m. They are fairly straight and the boles may be clear to 24m. Over-mature trees may be hollow.

## The timber
For practical purposes, both mersawa and krabak are similar, but the colour of the heartwood varies slightly with the species. The sapwood is not differentiated in freshly sawn timber but in course of time it is attacked by blue-stain fungi and becomes fairly distinct or distinct. The colour of the heartwood is yellowish-brown, occasionally with a rose tinge, darkening on exposure to a straw-brown. The wood is without lustre and featureless except for a mild form of silver figure caused by the rather prominent rays, and a subtle stripe figure on radial surfaces. The grain is straight or slightly interlocked, and the texture is moderately coarse and even. Moderately hard, and moderately heavy, the wood varies in weight, according to the species from 512 to 736 kg/m$^3$, generally averaging about 640 kg/m$^3$ when dried.

## Drying
All the species tend to dry very slowly from the green, and moisture is extracted from the centres of thick stock with some difficulty, but degrade is usually confined to slight distortion. Air dried wood that has to be kiln dried is accordingly non-uniform in its moisture spread, and much care is needed in assessing the initial moisture content on which to base the kiln schedule, and on the final moisture content if wet patches are to be avoided.

## Strength
Australian sources suggest the strength properties of mersawa are similar to those of abura (*Mitragyna ciliata*) and agba (*Gossweilerodendron balsamiferum*).

394

## Durability
Malaysian sources classify mersawa as being very durable, but tests elsewhere classify both mersawa and krabak as moderately durable.

## Working qualities
Both mersawa and krabak vary in their working qualities because of the silica content of the wood, but in general, while they are relatively easy to work, the blunting effect on cutting edges is often severe, especially on saw teeth, which should be tipped with hard metal for the most satisfactory results, and should have a long tooth pitch when a band mill is employed in conversion. The wood can be planed and moulded to a reasonable finish, but a reduction of cutting angle to 20° is helpful, particularly with quarter-sawn material which usually tends to tear. It nails and screws quite well, and takes a good polish.

## Uses
Joinery, flooring, general construction; locally used for boat planking and furniture. Suitable for plywood of the utility class such as for tea chests.

# NYATOH

### Species of Sapotaceae

## Other names
njatuh (Indonesia), or padang (UK), are trade names given to certain species of the Sapotaceae family which produce light to medium-weight timbers of similar colour, density, and properties. The following genera and species are usually employed in this product, those generally dominating commercial supplies are underlined.
*Palaquium maingayi, P. rostratum, P. xanthochymum, P. clarkeanum, P. cryptocariifolium, P. gutta, P. herveyi, P. hexandrum, P. hispidum, P. microphyllum, P. obovatum, P. semaram, P. walsurifolium, Payena maingayii, P. dasphylla, P. lanceolata, P. obscura. Ganua motleyana.*

See also **bitis** described elsewhere in this book.

### General characteristics
The sapwood of nyatoh timbers is lighter-coloured, but only moderately differentiated from the heartwood, which is deep pink-brown or red-brown. Planed surfaces generally without lustre; sometimes with a slight streaky figure on quarter-sawn surfaces. The grain is interlocked, and the texture moderately fine and even, and the wood weighs on average 720 kg/m³ when dried.

### Drying
The wood dries at a moderate rate, with only a slight tendency to distortion, and a moderate tendency for surface checks to develop, more so around knots.

### Strength
The strength properties are similar to those of closely related makoré (*Tieghemella heckelii*) of West Africa.

### Durability
Moderately durable.

### Working qualities
Works reasonably well, there is some tendency for saws and cutters to blunt fairly rapidly, and for the grain to tear in planing and moulding quarter-sawn material. A reduction of cutting angle to 20° is beneficial. The wood gives excellent results in staining and polishing; it glues well, but tends to split when nailed.

### Uses
Furniture, cabinet-making, patterns (the wood has low shrinkage), interior building construction, high-class joinery, veneer, plywood, doors and shingles.


## PENARAHAN

### Species of Myristicaceae

Various species of the Myristicaceae family occur in Malaysia and Indonesia, and the Malayan light hardwood known as penarahan may be drawn from some 20 different species, all however, with similar properties and characteristics. The principal species dominating commercial supplies are given below.

*Gymnacranthera bancana, Myristica gigantea* and *M. lowiana.*

### General characteristics
The sapwood is not differentiated from the heartwood by colour; the wood is pale brown or pinkish-brown, and in many of the species there is an occasional distinctive core, some 150mm wide or more, the wood of which is deep purple-red or chocolate-red in colour. Irregularly spaced, terminal parenchyma bands produce a growth-ring figure on tangential surfaces. The grain is almost straight, and the texture moderately fine, but not very even due to the relatively widely dispersed vessels. The wood varies in weight according to the species, from 480 to 736 kg/m³ but averages about 590 kg/m³ when dried.

### Uses
Penarahan is non-durable, and provided the dark-coloured core is excluded in conversion (apart from colour, this usually contains deposits which may reduce the overall machining properties), the timber is suitable for interior joinery and light construction.

## PERUPOK

*Lophopetalum* spp.                    Family: Celastraceae

Perupok is produced in Malaysia from species of *Lophopetalum*, principally from *L. javanicum* and *L. subovatum*, but the following may be included, *L. pachyphyllum, L. pallidum, L. wighteanum,* and *L. maingayi*; this latter species is heavier and denser than the general run of commercial perupok.

### General characteristics
The wood is straw-coloured when freshly cut, turning a light brown on exposure. There is no differentiation between sapwood and heartwood by colour. Planed surfaces have a medium lustre, and tangential surfaces show a feathery tracery produced by wood parenchyma bands. The grain is interlocked,

397

and the texture moderately fine and even. The usual, average weight of commercial parcels of perupok is between 480 and 640 kg/m³ when dried.

The wood is reputed to dry without excessive degrade, to work fairly easily, and to take a good polish. It is not durable in contact with the ground.

## Uses
Interior joinery and light construction.

# PULAI

*Alstonia* spp.                                Family: Apocynaceae

Malaysian pulai is produced mainly from *Alstonia angustiloba*, and *A. spathulata*, but a third species, *A. scholaris* may also be included in commercial shipments.

## The timber
There is no difference in colour between sapwood and heartwood, the wood being creamy-white or light-yellowish-brown, with occasional zig-zag markings on tangential surfaces due to wood parenchyma bands. Latex traces may appear on tangential surfaces as lens-shaped scars arranged in whorls similar to jelutong (*Dyera costulata*).

The grain is interlocked, and the texture moderately fine to moderately coarse. It weighs 368 to 480 kg/m³, averaging about 464 kg/m³ when dried.

## Drying
The timber dries rapidly with only a slight tendency for checks and splits to develop.

## Strength
Pulai has low strength properties; similar to obeche (*Triplochiton scleroxylon*) in most categories, but rather lower in shear strength, hardness and toughness.

## Durability
Perishable.

## Working qualities

Works easily with hand and machine tools, but due to the soft nature of the wood, cutting edges must be kept sharpened if crumbling of the grain is to be avoided. It glues, stains and polishes well, but the appearance of the wood is often marred by the latex traces.

## Uses

The timber has a good reputation for stability in service, and is therefore suitable for pattern-making. It is also used for matches, tea boxes, crates, carving, plywood and carpentry.

# PUNAH

*Tetramerista glabra* Miq.             Family : Marcgraviaceae

## Other names

entuyut (Sarawak and Brunei) ; larut, kuantan (Malaysia).

## Distribution

Malaysia, Sarawak, Brunei, East Kalimantan, Sumatra and some other islands of Indonesia.

## The tree

A large evergreen tree, attaining a height of 36m and a diameter of 1.2m or more. The boles are generally straight and clear for about 15m above the fluted base.

## The timber

The sapwood is about 38mm wide, and not defined when freshly cut but fairly clearly demarcated in dried timber. The heartwood is straw-coloured or yellowish pink, weathering to a pinkish-brown with splashes of an orange-brown tinge. Planed surfaces are without lustre or figure, but the broader rays of the two types present are conspicuous on quarter-sawn surfaces. The grain is straight or shallowly interlocked or spiral, and the texture is coarse but even. Moderately hard and moderately heavy, the wood weighs about 720 kg/m$^3$ when dried. The wood contains saponin and will produce a lather if a few thin slivers are placed in warm water and vigorously shaken.

## Drying
It dries easily but is liable to cupping and end splitting unless carefully handled.

## Strength
In strength properties it is said to be 30 per cent stiffer than English oak (*Quercus* spp.), and 15 per cent to 20 per cent stronger in compression and bending, but is softer and less resistant to shock.

## Durability
Probably moderately durable.

## Working qualities
It is said to saw easily with slight gumming of the saws. It is capable of a good finish after sanding, and polishes well after suitable filling. Tends to split when nailed.

## Uses
Interior joinery, structural work not in contact with the ground, boxes and crates, carpentry and window sills.

# RAMIN

*Gonystylus macrophyllum* (Miq.)      Family : Gonystylaceae
Airy Shaw

## Other names
melawis (Malaysia) ; ramin telur (Sarawak).

## Distribution
Sarawak and Malaysia, where it is common in the fresh-water swamps, particularly on the west coast.

## The tree
A medium-sized tree, attaining a height of about 24m with a straight, clear, unbuttressed bole about 18m long and 0.6m in diameter.

## The timber
The sapwood is not differentiated from the heartwood by colour, but is usually from 38mm to 50mm wide. The heartwood is a uniform pale straw colour; planed surfaces are practically without lustre or figure. The grain is straight to shallowly interlocked, and the texture is moderately fine and even. The wood weighs about 670 kg/m$^3$ when dried.

Note: melawis is the product of three species of *Gonystylus* in Malaysia, *G. macrophyllum*, *G. affinis*, and *G. confusus*. The timber has been exported from Malaysia under this trade name, but essentially, there is little difference between ramin and melawis, since the dominant species making up the latter is *G. macrophyllum*. The other two species however, are a little heavier, averaging 705 kg/m$^3$.

## Drying
The timber dries reasonably well, but with a tendency for end splitting and surface checking to occur. The wood is prone to staining, and must therefore be dipped immediately following conversion. Good, clean, air dry stock is easily kiln dried, without undue degrade.

## Strength
Ramin is very similar to beech (*Fagus sylvatica*) in its general strength properties, but it is less tough and hard, and is weaker in shear, and unlike beech is unsatisfactory for bent work. It is however, much stronger in compression than beech.

## Durability
Non-durable.

## Working qualities
The timber works fairly easily with both hand and machine tools, but in planing and moulding there is a tendency for the grain to tear occasionally in quarter-sawn material, and it is sometimes advisable to reduce the cutting angle to 20°. The timber turns reasonably well, and it can be glued, painted, stained and polished satisfactorily. It is liable to split if nailed near the edges.

## Uses
Furniture, interior joinery, handles for non-striking tools, turnery, toys, panelling, veneer, flooring, carving, shop-fittings and small mouldings.

# RENGAS

## Species of Anacardiaceae

Various genera and species of the Anacardiaceae family occur in Malaysia the Philippines, and New Guinea. Those producing Malaysian rengas are given below, the types usually dominating the production being underlined.
*Melanorrhoea curtisii, M. torquata* King., *M. wallichii* Hook. f., *M. aptera, M. malayana* and *M. woodsiana.*
*Gluta elegans, G. renghas, G. wrayi,*
*Melanochyla auriculata, M. bracteata, M. kunstleri, M. rugosa.*

## General characteristics
The sapwood is well defined and sharply contrasted; the heartwood which is blood-red-coloured, and streaky, standing out from the pinkish-brown sapwood. An unusual but characteristic feature of rengas is the presence of horizontal canals in the rays. These exude a colourless sap which darkens on exposure and forms blackish blotches on the longitudinal surfaces of the wood. Accordingly, both the sapwood and heartwood of freshly sawn wood is invariably marked with dark blotches, but after the wood is dried, and the surfaces are planed, the blotches disappear.
The wood has an interlocked grain, and a texture that varies with the species from moderately fine to moderately coarse. The weight is also variable, ranging from 672 kg/m$^3$ to 990 kg/m$^3$ when dried.

## Drying
The wood is reported to dry easily, and with little degrade, and is said to have a very low to low shrinkage.

## Strength
The average strength of the species of *Melanorrhoea* (the main species making up rengas) is roughly similar to European elm (*Ulmus procera*), but rengas is some 50 per cent harder, and

about 30 per cent stronger in compression than elm, but is not quite as tough.

## Durability
Moderately durable to durable.

## Working qualities
Rengas works fairly easily, but requires some care in finishing to avoid grain tearing. The wood takes stains and polish satisfactorily.

## Uses
An attractive wood used for furniture, cabinet-making and ornamental work.

# RESAK

*Vatica* and *Cotylelobium* spp.          Family: Dipterocarpaceae

A number of species of *Vatica* and *Cotylelobium* are common throughout south-east Asia, and several may be combined in the production of commercial resak. The following are some of the species frequently used in Malaysia, those usually dominating commercial supplies being underlined.
*Vatica cinerea, V. cuspidata, V. odorata, V. stapfiana, V. bancana, V. bella, V. flavida, V. havilandii,* and others, and *Cotylelobium malayanum.*

## General characteristics
The sapwood is moderately well defined, and the colour of the heartwood is yellow-brown or light-brown, sometimes with an olive tinge when freshly sawn, weathering to chocolate-brown. The fairly large rays produce a silver-grain figure on quarter-sawn surfaces. The grain is shallowly interlocked and the texture is fine and even, and the wood weighs between 656 kg/m$^3$ to 960 kg/m$^3$ when dried, depending on the species, the average weight is about 768 kg/m$^3$.

It is said to be rather difficult to work, partly because of the hard, to very hard nature of the wood, and partly because of the intercellular resin canals, both of which combine to blunt cutting edges fairly quickly. The heavier varieties are said to be very durable, the others only moderately durable.

403

## Uses
Generally considered a heavy constructional timber, and suitable for marine uses.

## SELANGAN BATU and BALAU

*Shorea* spp.                                        Family: Dipterocarpaceae

The hard, heavy timber species of *Shorea* occurring in South East Asia are grouped under common trade names peculiar to the area. Thus **selangan batu** (hard selangan) is the name used in Sabah, Brunei, and Sarawak, while **balau** is used to describe the heavy Malaysian species.

Formerly, selangan batu (as a group), included heavy timbers of the genus *Hopea*, but these are now separated into a distinguishing description as **giam**, or as **merawan**, according to the area of origin.

### selangan batu
This is produced by several species of *Shorea*, but principally by *S. laevis* Ridl., and *S. seminis* V.Sl.

Selangan batu is recognised as two separate types in Sabah, selangan batu No 1 timber with an air dry weight of 881 kg/m$^3$ or more and,
selangan batu No 2 timber with an air dry weight less than 881 kg/m$^3$.

Selangan batu No 1, is roughly the equivalent of Malaysian balau.

Note: Timber produced in Sarawak and Brunei from *Shorea albida* Sym., is known as **alan**, with alternative names as follows,
meraka alan, red selangan (in part), selangan merah (in part) (Sarawak); seringawan (Brunei). These weigh 880 kg/m$^3$, the lighter weight timber of this species is selected out and classed as light red meranti.

**Red selangtan batu**, or selangan batu merah, is the product of several *Shorea* species but principally *S. guiso* Bl., the name

being applied in Sabah as distinct from selangan batu No 2.
The wood weighs about 850 kg/m³ air dry.

## Balau
Balau is generally separated into two types in Malaysia, ie,
balau and red balau.
Balau is produced mainly from *S. atrinervosa, S. elliptica,
S. foxworthyi, S. glauca, S. laevis, S. maxwelliana,* and *S.
submontana.*
Red balau is produced mainly from *S. guiso, S. kunstleri,
S. collina,* and *S. ochrophloia.*

### General characteristics

**Selangan batu** is a yellowish-brown timber with interlocked
grain, and a coarse, but even texture. According to species, durable to very durable.
**Balau** is a yellowish-brown, brown, or reddish-brown timber
with interlocked grain, and a moderately fine and even texture.
It is classified in Malaysia as being very durable.
**Red balau** is a purplish-red or dark red-brown timber, with an
interlocked grain, and a coarse but even texture. It is classified
in Malaysia as being less durable than balau, ie moderately
durable.
**Red selangan batu** (Sabah), and alan, or **meraka alan** (Sarawak
and Brunei) are approximately equal in colour and characteristics to **red balau.**

### Average weights (air dry)

| | |
|---|---|
| balau and selangan batu No 1 | 881 to 980 kg/m³ |
| red balau | 800 to 880 kg/m³ |
| selangan batu No 2 | 850 kg/m³ |
| alan | 850 to 880 kg/m³ |
| red selangan batu | 850 kg/m³ |

### Average strength properties
All the species mentioned have hard, heavy, and strong timbers
with strength properties similar to those of greenheart.

### Uses
All the species are suitable for heavy structural work, bridge and
wharf construction, sleepers, flooring, and boat framing, but the

higher durability of balau and selangan batu renders them more suitable for adverse conditions of use.

## SEPETIR and SWAMP SEPETIR

*Sindora* spp.                                        Family : Leguminosae

**Sepetir** was formerly recommended as the timber name for species of *Sindora* and the closely related *Pseudosindora palustris*. Since the timber of the latter species differs from that of species of *Sindora*, it is proposed by the British Standards Institution that *Pseudosindora palustris* should be known as **swamp sepetir**.

### Other names
makata (Thailand) for sepetir.
sepetir paya (Sarawak) for swamp sepetir.

### Distribution
Of the various species of *Sindora* occurring in Malaysia, two species, *S. coriacea* and *S. echinocalyx* dominate commercial supplies of sepetir from that area, but *S. siamensis, S. velutina* and *S. wallichii* may also be included. *Pseudosindora palustris*, which was formerly classified as *Copaifera palustris*, occurs in Sarawak and Brunei in the freshwater swamps, in association with ramin.

### The tree
*Sindora* spp., are generally large trees, some 45m in height, and a diameter of about 1.2m in favourable conditions, but more commonly the diameter is about 0.75m and the boles up to 12m to 15m long. *Pseudosindora palustris* is a smaller tree, which usually produces a straight, cylindrical bole about 0.6m in diameter.

### The timber
There are slight, but subtle differences between the timbers of sepetir and swamp sepetir sufficient to affect the end uses which, for both types are the same. The various species of *Sindora* are typically brownish coloured woods, with some variation in weight, and possessing axial intercellular canals

in the terminal layers of wood parenchyma which secrete oil when freshly cut. Swamp sepetir on the other hand, has wood of a typically reddish-brown colour, more uniform both in colour and weight because it is the product of a single species, and without the vertical canals.

Both types have a wide sapwood, that in sepetir varying from 75mm to 300mm wide, and in swamp sepetir commonly 100mm wide. Since this is very susceptible to beetle attack, its inclusion in many forms of end use is risky, and sapwood therefore is best excluded. The heartwood of *Sindora* is pinkish-brown, golden brown, or red-brown when first cut, darkening to darker shades, and often streaked with darker-coloured layers, suggesting walnut (*Juglans regia*). Planed surfaces are without significant lustre; the grain is shallowly interlocked, and the texture moderately fine and even, or moderately coarse. A growth-ring figure appears on tangential surfaces due to the irregularly spaced terminal parenchyma bands, while quarter-sawn surfaces have a stripe figure. The wood has a greasy feel, and is moderately hard and moderately heavy on average, although the weight of the various species range from 640 kg/m³ to 830 kg/m³ with an average for commercial parcels of about 672 kg/m³ when dried.

The heartwood of swamp sepetir is pale pink with pale brown veining when freshly cut, darkening on exposure to a rich reddish-brown, relieved by the darker veining. A growth-ring figure appears on tangential surfaces, but the wood is of a plainer appearance than Malaysian sepetir. The grain is usually straight, and the texture moderately fine and even, but occasionally moderately coarse. The wood weighs about 670 kg/m³ when dried.

### Drying
Both types dry slowly but very well, there being only a slight tendency to distortion, but there is a tendency for end splitting to occur.

### Strength
Both types are similar to English oak (*Quercus* spp.) in shock resistance and toughness, but they are generally superior to oak in hardness, shear, stiffness, and in bending and compression.

## Durability
The heartwood is durable, but the sapwood should be considered perishable.

## Working qualities
Sepetir is rather difficult to saw and plane due to the resin in the wood which tends to make for sticky surfaces and to make saws gum up fairly quickly. The wood can be polished satisfactorily, the amount of resin apparently being insufficient to interfere seriously with this operation. It is rather difficult to prevent the wood splitting when nailed, and it should be pre-bored for this purpose.
Swamp sepetir is rather more easily worked because of the absence of resin.

## Uses
Both types of wood are suitable for a wide variety of uses, including furniture, cabinet-making, joinery, panelling, musical instruments, sporting goods and veneer. The more highly decorative of these uses is better served by selection of suitable sepetir, the more subdued appearance of swamp sepetir being applied to less decorative forms.

# SESENDOK

*Endospermum malaccense* Muell. Arg.   Family: Euphorbiaceae

## Other names
sendok-sendok.

## General characteristics
The sapwood is not differentiated from the heartwood; the wood is bright yellow with a greenish tinge when freshly cut, turning to a straw colour on exposure. Planed surfaces are lustrous, and may show concentric markings due to occasional layers of darker tissue. The grain is spiral, interlocked, or wavy, and the texture coarse but even. The wood weighs about 528 kg/m$^3$ when dried.
The timber dries fairly well, but with a tendency to distort. It is soft, and easy to saw, but the variable grain may give some trouble in planing and moulding. It takes a good polish, but the

pores, although relatively few in number, are moderately large in size, and filling is required to obtain the best finish.

## Uses
Sesendok is not considered durable in Malaysia, but since this reference is concerned more with resistance to termites rather than to decay, it is probably moderately durable when in contact with the ground. The timber is suitable for joinery and light construction.

# SIMPOH

Species of *Dillenia*                                  Family : Dilleniaceae

Several species of *Dillenia* occur in Malaysia, and the following produce a medium hardwood known as simpoh.
*Dillenia eximia, D. reticulata, D. obvobata, D. ovata, D. pulchella* and *D. indica.*

## General characteristics
The sapwood is light red-brown in colour, not sharply demarcated from the heartwood which is red-brown or dark red-brown, often with a purple tinge. The broad rays produce a silver-grain figure on radial surfaces ; the grain is interlocked, and the texture moderately coarse and even. The wood weighs about 736 kg/m³ when dried. Siliceous deposits generally occur in vessel and fibre walls.

## Drying
The timber is reputed to have high to very high shrinkage properties, and coupled with the broad rays, there is a distinct tendency for surface checking and end splitting to occur unless care is taken.

## Strength
Similar to elm (*Ulmus procera*) in most strength categories, but slightly inferior to that timber in shear strength, hardness and toughness.

## Durability
Non-durable.

## Working qualities
Although somewhat difficult to work, the dulling effect on cutters and saws being relatively high, the wood is capable of a good finish.

## Uses
Panelling, furniture, light interior construction and plywood.

# TEAK

*Tectona grandis* L.f.                    Family : Verbenaceae

Teak is one of the world's best known and valued timbers. It occurs in many tropical areas, either as an indigenous species or as an introduced one. It is indigenous to India, Burma, Thailand, and to the former Indo-China, ie Cambodia, Laos, and Vietnam, and in Indonesia, particularly Java, although strictly speaking, teak was introduced into Java originally, and was planted near temples and shrines. Over several centuries these original plantings have extended and developed into almost pure timber stands over fairly extensive areas. Teak is not normally gregarious in its natural habitat, usually being found in mixed deciduous forests.

Teak has also been planted extensively in many other tropical regions including West Africa, the Philippines, Tropical America and the West Indies.

For an account of plantation-grown teak, see the companion Red Booklet, 'Timbers of Central America and the Caribbean'.

## The tree
In its natural habitat, teak is often a large tree, with a clean, cylindrical bole above a fluted or often buttressed base. It can vary considerably in height, girth and form, according to locality, particularly in regard to soil conditions. On clay soils it does not do well, and in the drier and hotter regions, the trees often have much shorter and more fluted stems, with more twisting and branching. On favourable sites, it can reach a height of 39m to 45m and a diameter of 1.5m with a clear bole of 10m up to 24m.

410

## The timber
The sapwood is yellowish or whitish in colour, and sharply defined from the heartwood, which is golden-brown, sometimes figured with darker markings, due to dark-coloured zones of initial parenchyma. Growth rings are distinct, but they vary considerably from a band of two or three layers of large pores to a few scattered large pores not forming a definite band. The dark-coloured parenchyma initiating the seasons growth likewise varies in its width and presence. Teak from some localities may display not only annually occurring growth rings, but also occasional false rings, with very little initial dark parenchyma, with the result that much teak, especially from Burma, is relatively uniform in colour but with only very narrow lines of darker colour marking the annual growth appearing on side grain. Indian teak, especially from the Malabar Coast, (with its heavy annual rainfall of some 3000mm) is usually more handsomely marked. The wood has an oily feel, and a strong odour reminiscent of old leather when freshly cut, but after drying much of the odour is lost, but the wood retains its oily feel.

Teak darkens in colour on exposure; it has a fairly straight grain, sometimes irregular, and a coarse, uneven texture. The average weight of Burma teak is about 640 kg/m$^3$ when dried.

## Drying
Teak from ring-girdled trees air dries easily but slowly. but timber from ungirdled, green trees requires a lot of initial care in order to avoid rapid drying which is liable to cause checking, end splitting and warping to develop. Teak presents no serious problems in kiln drying from the air dry condition, except in the assessment of initial and final moisture contents. While drying defects are usually minimal, there is usually considerable variation in the drying rate of individual pieces, and moisture content differences can be great.

There is often a loss of colour in some pieces in kiln drying, but the colour is quickly regained when the wood is exposed to light.

## Strength
The general strength properties of teak are about the same as those of mengkulang (*Heritiera* spp.). It compares favourably

with English oak (*Quercus* spp.) in most strength categories, but is rather weaker in shear strength and toughness.

## Durability
Very durable.

## Working qualities
Although variable, the wood can be worked with moderate ease with both hand and machine tools. There is a moderate to severe dulling of cutting edges, but if these are kept sharpened, the wood finishes well. It takes nails and screws fairly well, and glues satisfactorily on freshly machined or sanded surfaces. It can be varnished or polished satisfactorily.

## Uses
Shipbuilding, decking, planking, deck-houses, bulwarks, furniture, cabinet-making, interior fittings and panelling, out-door building and furniture, laboratory benches and equipment, acid vats, weather doors, plywood and decorative veneer.

# TEMBUSU

*Fagraea fragrans* Roxb.                    Family : Loganiaceae
syn. *F. cochinchinensis* (Lour.) A. Chev.

There are at least two other species of *Fagraea* occurring in South East Asia, *F. gigantea* Ridl. and *F. racemosa* Jack. The former species known in Sarawak as tembusu hutan, is a large tree, and the timber is very similar in appearance and properties to tembusu and can be used for the same purposes. It is however lighter in weight, averaging about 672 kg/m$^3$. *F. racemosa* is a small tree and the timber is of little commercial value.

## Other names
anan, yellowheart (Burma) ; temasuk (Sabah) ; tembusu pedang (Sarawak and Brunei) ; lemesu, meraing, reriang (Malaysia).

## Distribution
Although widely distributed throughout the area, it is nowhere abundant.

## The tree
A fairly large evergreen tree, attaining a height of 30m and a maximum diameter of 0.75m. The bole is fluted and irregular and may be clear of branches up to 18m.

## The timber
Sapwood and heartwood are not usually distinct, although the former may be slightly lighter in colour, pale yellow, and fairly wide. The heartwood is whitish to yellowish-brown when freshly cut, darkening on exposure to a light orange-brown Planed surfaces are lustrous; tangential surfaces have light coloured zig-zag markings produced by parenchyma. The texture varies from moderately fine to moderately coarse, and the grain is slightly interlocked and often wavy. The wood varies in weight from 656 kg/m$^3$ to 993 kg/m$^3$ but averages 816 kg/m$^3$ when dried.

## Drying
Dries slowly with a tendency to surface check.

## Strength
No data are available.

## Durability
Very durable.

## Working qualities
Reported to be similar to beech in working qualities but with rather more dulling effect on cutting edges. It carves easily, takes a good finish, although care is needed with irregular-grained material. Takes a high polish.

## Uses
Tembusu is used in the countries of origin for furniture, cabinet-making, carving, chopping blocks, doors, sills, coffins and constructional work. It makes an attractive hard-wearing floor, and is said to be excellent for piling, bridge construction, boat building and for waggon bottoms.

413

# TERAP

## Species of Moraceae

Terap as produced in Malaysia, is obtained principally from *Artocarpus scortechinii*, but other species of the Moraceae family may be included, usually *Artocarpus elasticus*, *Partocarpus bracteatus* and *P. venenosus*.

### General characteristics

The sapwood generally is not differentiated from the heartwood which in some species is yellowish-brown with an orange tinge, or orange-brown (in which case, the sapwood is lighter in colour). Planed surfaces commonly show a stripe figure when quarter-sawn; the grain is interlocked, and the texture coarse but even. Terap weighs about 496 kg/m³ when dried.

The timber dries fairly rapidly with a slight tendency for surface checks and end splits to develop. It is soft and easy to saw, but cutting edges are likely to become dulled fairly quickly due to the siliceous linings to the vessels and fibres common to most species of *Artocarpus*. The interlocked grain is liable to tear in planing and moulding, and a cutting angle of 20° is suggested. The wood polishes quite well after suitable filling.

### Uses

Joinery and light construction.

# TERENTANG

*Campnosperma* spp.                    Family: Anacardiaceae

Malaysian terentang is principally the product of *Campnosperma auriculata* Hook f., but other species may be included, eg *C. coriacea*, *C. montana*, *C. macrophylla* and *C. zeylanicum*.

### General characteristics

Sapwood not differentiated from heartwood by colour; the wood is greyish-pink or mauve-grey, and is rather plain looking with no lustre or figure, although longitudinal surfaces are mildly speckled by the darker-coloured rays. The grain is

interlocked, and the texture fine and even. The wood is very soft and light in weight, about 430 kg/m³ when dried.
The wood is said to dry rapidly and well, with little degrade, and is easy to work and finish. It is non-durable.

## Uses
Match boxes, packing cases, light construction, and veneer from selected logs.

# WHITE SERAYA

*Parashorea malaanonan* Merr.  Family: Dipterocarpaceae
syn *P. plicata* Brandis
and *P. tomentella*
(Sym) W. Meijer syn (*P. malaanonan* Merr.
var *tomentella* Sym.

## Other names
urat mata (Sabah).
White seraya is not the equivalent of Malaysian white meranti. It is the product of one species and its variant, and is therefore more uniform in its character.

## Distribution
Confined to Sabah in South East Asia, but occurs also in the Philippines where the timber is known as bagtikan.

## The tree
A large tree generally 36m in height with a diameter of 1.0m or more. The bole is long, straight, and cylindrical above the strong buttress.

## The timber
The sapwood is not clearly defined from the heartwood which is straw-coloured to very pale brown, occasionally with a pinkish tinge. The grain is interlocked, sometimes spiral, and the texture is rather coarse. The interlocked grain gives rise to a conspicuous ribbon figure on radial surfaces. The wood is moderately hard and moderately heavy, and weighs about 530 kg/m³ when dried. Large trees may contain ring shakes and spongy, brittle hearts.

415

## Drying
Although drying rapidly and quite well, it requires more care than red seraya. (See under **Meranti** for *Shorea* spp. producing red seraya).

## Strength
White seraya is comparable to American mahogany, but is slightly superior to that timber in maximum bending, stiffness and compression.

## Durability
Non-durable.

## Working qualities
The wood works reasonably well with both hand and machine tools. There is usually little dulling effect on cutting edges, but these should be thin and sharp in order to reduce the slight tendency for woolliness and grain tearing during planing and moulding. Nails, screws and glues, reasonably well, and stains and polishes satisfactorily with suitable filling.

## Uses
Joinery, carpentry, plywood, flooring (light pedestrian), and when specially selected, as decking in boats and ships.

# PART II  SOFTWOODS

## KAURI, EAST INDIAN

*Agathis alba* Foxw.  Family: Araucariaceae

## Other names
Borneo, Sarawak, Malaysian kauri (UK); bindang, bendang, (Sarawak); menghilan (Sabah); damar minyak (Malaysia).

## Distribution
East Indies and Malaysia.

## The tree
The tree varies in height according to the locality in which it

grows; it is reported to be a very large tree in Sabah, with a height of 60m and a diameter of 2.7m but generally the diameter is smaller being about 1.8m. In Malaysia and Sarawak the tree is generally smaller, with a diameter of around 0.75m. The bole is straight and cylindrical but since it tends to retain its branches, the timber is inclined to be knotty.

## The timber
Sapwood and heartwood are not clearly defined. The heartwood is light yellow or straw-coloured, often with a pink tinge, weathering to a golden-brown or light pinkish-brown; occasionally with a darker-coloured core which is distinct from the outer layers of wood. Planed surfaces are fairly lustrous and tangential faces generally have a growth-ring figure. The wood is non-resinous, and without odour. It has a fine and even texture, and is generally straight-grained, and weighs about 480 kg/m$^3$ when dried.

## Drying
Dries easily and well, with a slight tendency to distort, and to split in the vicinity of knots.

## Strength
The strength properties are similar to those of Parana pine (*Araucaria angustifolia*)

## Durability
Non-durable.

## Working qualities
Easily worked and finished. It turns well and takes a good polish.

## Uses
Interior joinery, pattern making, household utensils, kitchen furniture.

## SEMPILOR

*Dacrydium elatum* Wall.                  Family: Podocarpaceae

## Other names
malor (Sarawak and Sabah).

417

## Distribution
Sarawak and Sabah.

## The tree
A moderately large tree, occasionally reaching a height of 30m and a diameter of 0.9m but usually smaller.

## The timber
There is little difference between sapwood and heartwood in colour, the wood generally being light brown to pinkish-yellow, It is similar to East Indian kauri, with straight grain and fine, even texture, but is harder, and slightly heavier, weighing about 550 kg/m$^3$ when dried.

## Drying
No information is available.

## Strength
No information is available, but its strength is probably similar to that of Parana pine.

## Durability
Non-durable.

## Working qualities
Works and saws reasonably well, with some dulling effect on cutting edges. It is capable of a good finish in planing and moulding, and takes glue and polish quite well.

## Uses
Interior joinery, light construction not in contact with the ground, boxes and crates, flooring, and is said to produce good utility veneer.

# USE GUIDE FOR SOUTH EAST ASIAN TIMBERS

## ACID AND DYE VATS

bangkirai

teak

## BOAT AND SHIP CONSTRUCTION

### Decking
bangkirai
belian
bintangor
chengal
melunak

meranti gerutu
teak
tembusu
white meranti
white seraya (selected)

### Framing
balau
melunak
selangan batu

teak
tembusu

### Keels and stems
belian
chengal

giam
tembusu

### Masts and spars
bintangor

### Planking
melunak
mersawa/krabak
teak

tembusu
white meranti

### Superstructures
bintangor
katon
melunak
mengkulang

meranti gerutu
meranti, dark red
teak
tembusu

## BOXES AND CRATES

binuang
durian
geronggang
kedondong (canarium)

pulai
punah
sempilor
terentang

# CONSTRUCTION

## Heavy

balau
bangkirai
belian
bitis
chengal
giam
kapur
kelat
kempas
keranji

keruing
mempisang
merbau
merpau
merawan
mersawa/krabak
resak
selangan batu
tembusu
tualang

## Light

bintangor
binuang
durian
jelutong
katon
kauri
kedondong (canarium)
keledang
keruing
kungkur
machang
medang
mempisang
mengkulang

meranti, dark red
meranti gerutu
meranti/seraya, light red
meranti, yellow
nyatoh
penarahan
perupok
punah
sempilor
sesendok
simpoh
terap
terentang

# DOORS

katon
medang
melunak
meranti (all types)

nyatoh
teak
tembusu
white seraya

# FANCY GOODS

amboyna

rengas

420

# FLOORING

balau
bangkirai (heavy duty and
    acid resisting)
bintangor
chengal
kapur
keruing
melunak
mengkulang
meranti (all types)

meranti gerutu
merbau
mersawa/krabak
ramin
sempilor
selangan batu
teak
tembusu
white seraya

# FURNITURE AND CABINET MAKING

amboyna
bintangor
katon
kauri (kitchen)
keledang
medang
melunak
mempisang

meranti
nyatoh
ramin
rengas
sepetir
simpoh
teak
tembusu

# JOINERY

**High-class**
bintangor
katon
kelat
medang
melunak
mempisang
mengkulang
meranti (all types)
meranti gerutu
merawan

mersawa/krabak
nyatoh
ramin
sepetir
simpoh
teak
tembusu
terap
white seraya

**Utility**
binuang
geronggang
kapur
kauri
kedondong (canarium)

keruing
kungkur
machang
medang
meranti (all types)

421

**Joinery  Utility** *continued*

| | |
|---|---|
| meranti gerutu | ramin |
| penarahan | sempilor |
| perupok | sesendok |
| pulai | swamp sepetir |
| punah | terentang |

## MARINE PILING AND CONSTRUCTION

**Under water**

**(a) Teredo infested waters**

| | |
|---|---|
| belian | keranji |
| keledang | resak |

**(b) Non-Teredo waters, in addition to above,**

| | |
|---|---|
| balau | merpau |
| bangkirai | selangan batu |
| chengal | tembusu |
| kapur | |

**Above water**

**(a) docks, wharves, bridges, etc.**

| | |
|---|---|
| balau | keruing |
| bangkirai | merawan |
| belian | merbau |
| bintangor | merpau |
| chengal | resak |
| giam | selangan batu |
| kapur | teak |
| kempas | tembusu |
| keledang | tualang |
| keranji | |

**(b) decking**

| | |
|---|---|
| bangkirai | merbau |
| belian | merpau |
| chengal | teak |
| giam | tembusu |
| keruing | |

## PATTERN MAKING

| | |
|---|---|
| jelutong | nyatoh |
| kauri | pulai |

## SPORTS GOODS

bintangor (diving boards)   sepetir

## STAIR TREADS

| | |
|---|---|
| bintangor | sepetir |
| machang | teak |
| medang | tembusu |
| melunak | white meranti |
| nyatoh | |

## TERMITE RESISTANCE (HEARTWOOD)*

**Very resistant**

| | |
|---|---|
| balau | giam |
| belian | selangan batu |
| bitis | teak |
| chengal | tembusu |

**Resistant**

amboyna   merbau

**Moderately resistant**

| | |
|---|---|
| kelat | nyatoh |
| keledang | red balau |
| keranji | rengas |
| kungkur | resak |
| merawan | terap |

**Susceptible**

| | |
|---|---|
| bintangor | mempisang |
| binuang | mengkulang |
| durian | merpau |
| geronggang | meranti (all types) |
| jelutong | meranti gerutu |
| kapur | mersawa |
| kauri | penarahan |
| kedondong | perupok |
| kempas | pulai |
| keruing | punah |
| machang | ramin |
| medang | sempilor |
| melunak | sepetir |

sesendok  
simpoh  
terentang  

tualang  
white seraya  

*The above classification refers to resistance to attack by both subterranean and dry-wood termites. Where the resistance to either type of pest differs, the lower rating is given.

## TURNERY

amboyna  
bitis  

ramin  
teak  

## VEHICLE BODIES

keruing  
meranti gerutu  

tembusu  
white meranti  

## VENEER AND PLYWOOD

**Corestock**  
kempas  

tualang  

**Decorative**  
amboyna  
nyatoh  
ramin  

sepetir  
teak  
terentang (selected)  

**Utility (plywood, chip-baskets, small laminated items etc.)**  
kedondong  
medang  
mengkulang  
meranti  
mersawa/krabak  
nyatoh  
pulai  

ramin  
sempilor  
simpoh  
swamp sepetir  
teak  
white seraya  

## AMENABILITY OF HEARTWOOD TO PRESERVATIVE TREATMENT

**Extremely resistant**  
balau  
dark red meranti/seraya  
kapur  

kedondong  
light red meranti  
merbau

nyatoh
red balau
red selangan
selangan batu
swamp sepetir

teak
tembusu
white meranti
white seraya
yellow meranti

## Resistant
bintangor
light red seraya
mengkulang

merawan
punah
red meranti/seraya

## Moderately resistant
binuang
geronggang
perupok

sepetir
terentang
white meranti

## Permeable
jelutong
kempas
machang

ramin
sesendok

Note : Because of the large number of species often involved in the production of South East Asian timbers, particularly of the *Shorea* genus, there is inevitably a variation in their resistance to impregnation and accordingly some timbers are given under more than one heading. In case of doubt, the higher classification should be considered more appropriate.

# AMENABILITY OF HEARTWOOD TO PRESERVATIVE TREATMENT

The above classification refers to the ease with which a timber absorbs preservatives under both open-tank (non-pressure) and pressure treatments. Sapwood, although nearly always perishable, is usually much more permeable than heartwood; accordingly, the above classification refers to the relative resistance of heartwood to penetration.

### Extremely resistant
Timbers that absorb only a small amount of preservative even under long pressure treatments. They cannot be penetrated to an appreciable depth laterally, and only to a very small extent longitudinally.

### Resistant
Timbers difficult to impregnate under pressure and require a long period of treatment. It is often difficult to penetrate them laterally more than about 3mm to 6mm.
Incising is often used to obtain better treatment.

### Moderately resistant
Timbers that are fairly easy to treat, and it is usually possible to obtain a lateral penetration of the order of 6mm to 18mm in about 2-3 hours under pressure, or penetration of a large pro-portion of the vessels.

### Permeable
Timbers that can be penetrated completely under pressure without difficulty, and can usually be heavily impregnated by the open-tank process.

# REFERENCES

BOLZA, Eleanor and KEATING, W G. African timbers—the properties, uses and characteristics of 700 species. Melbourne, Australia, Division of Building Research. 1972.

BRAZIER, J. D. Notes on some Brazilian Timbers; 1975. Institute of Wood Science, London; One-day Seminar; Timbers of South America.

BRITISH STANDARDS INSTITUTION. Nomenclature of commercial timbers, including sources of supply. British Standard BS 881 & 589. London, BSI. 1974.

BUILDING RESEARCH ESTABLISHMENT. Handbook of hardwoods, revised by R. H. Farmer. London, HMSO. 1972.

BUILDING RESEARCH ESTABLISHMENT. A handbook of softwoods. BRE Report. London, HMSO. 2nd ed. 1977.

INSTITUTO BRASILEIRO DE DESENVOLVIMENTO FLO-RESTAL, Rio de Janeiro. Madeiras do Brasil. 1965.

KLOOT, N. H. and BOLZA, E. Properties of timbers imported into Australia. Australia. Division of Forest Products, Technological Paper 12. Melbourne, CSIRO. 1961.

KRYN, Jeannette M and FOBES, E W. Woods of Liberia. US. Forest Products Laboratory Report 2159. Madison, FPL. 1959.

MENON, P. K. B. Structure and identification of Malayan woods. Malaya. Forest Research Institute, Forest Record No. 25. Kuala Lumpur, Forest Department. 1967.

OKIGBO, L. Some Nigerian woods. 2nd edition. Lagos, Federal Ministry of Information. 1964.

PEARSON, R. S. and BROWN, H. P. Commercial timbers of India. Calcutta, Government of India, Central Publications Branch. 1932. 2 vols.

## REFERENCES *(cont)*

RECORD S. J. and HESS R. W. Timbers of the New World. 1943. Yale University Press, Oxford University Press, (New Haven, Oxford and London).

U.S. DEPARTMENT OF AGRICULTURE, Washington D.C. Agricultural Handbook No. 207. Present and Potential Commercial Timbers of the Caribbean 1971.

*Addenda*

# SOUTHERN ASIA
## BIJASAL

*Pterocarpus marsupium* Roxb.                    Family: Leguminosae

Other Names
bija, pisal, honne, vengai

### Distribution and description
A large, deciduous tree, widely distributed in India and Sri Lanka, especially in central and southern India. It is the source of an oily gum from which is produced Malabar kino, used in medicine.

### The timber
The sapwood is yellowish white in colour, fairly well-defined from the heartwood which is golden or golden-brown, often with darker streaks. Ripple marks are conspicuous on tangential surfaces and thin, wavy bands of parenchyma connect the pores. These are fairly large and are often filled with a dark gum. The wood is close-grained, hard and heavy, weighing on average 800 kg/m$^3$ when dry. It works and machines reasonably well and is capable of taking a fine polish. It is durable in respect of resistance to decay.

### Uses
Doors and window frames, posts and beams, agricultural implements vehicle bodies and furniture.

## POON

*Calophyllum* spp.                    Family: Guttiferae

Other Names
*Calophyllum tomentosum* Wight syn *C. elatum* Bedd. produces poon spar, nagari, pongu and surhoni.
*Calophyllum wightianum* Wall produces irae and sirapunme.
Note. *Calophyllum inophyllum* L. is also sold occasionally as poon, but due to its smaller pores it has a smoother, more compact character than the other species and for that reason is more commonly sold in India as bobbi or bobli.

## Distribution and description
Poon, or poon spar tree, is a tall, straight, relatively slender tree found particularly in the Western Ghats and adjoining hills, and in the north and south Kanara districts of India.

## The timber
The wood is similar in appearance to the related bintangor of Malaysia and beach calophyllum and pink touriga of Australia and Papua New Guinea. The sapwood is reddish-white in colour and the heartwood light reddish-brown with darker streaks. The wood has a rather coarse texture and a grain which may be interlocked, spiral or wavy. It is moderately hard and moderately heavy, weighing about 672 kg/m$^3$ when dry. Poon, of all species, requires care in seasoning in order to avoid warping and, although fairly easy to work, tends to produce a woolly surface in planing. The wood is moderately durable.

## Uses
Formerly, there existed a great demand for poon spars for native dhows but this is now much reduced, modern requirements generally being for bridge building and construction, and for plywood manufacture. Poon was used in the UK in the 1920's for joinery and other purposes as an alternative to mahogany.

# VELLOPINE

*Vateria indica* L.              Family: Dipterocarpaceae

## Other Names
vellapiney, white damar tree, dhupa, dhup maram, 'Malabar white pine.'

## Distribution and description
A large, handsome tree of the evergreen forests of the Western Ghats and extensively planted in India as an avenue tree. Also grows in Sri Lanka. It is a principal source of white damar resin used in spirit varnishes and nitrocellulose lacquer.

## The timber
The sapwood is whitish in colour, sometimes tinged with red or grey, and the heartwood is whitish-grey turning brownish on exposure.

The wood is soft and moderately heavy, weighing about 576 kg/m$^3$ when dry.

**Uses**
plywood, tea chests, coffins and packing cases.


## USE GUIDE FOR SOUTHERN ASIAN TIMBERS


BOAT AND SHIP CONSTRUCTION
**Masts and spars:** poon

CONSTRUCTION
**Heavy:** bijasal
**Light:** poon, vellopine

DOORS: bijasal

JOINERY
**High class:** poon
**Utility:** bijasal

MARINE PILING
**Non-teredo waters:** bijasal
**Above water,** docks, wharves, bridges, etc: bijasal, poon

VEHICLE BODIES: bijasal

VENEER AND PLYWOOD
**Utility:** poon, vellopine


## AMENABILITY TO PRESERVATIVE TREATMENT


**Resistant:** bijasal, poon

# INDEXES

## INDEX: 1. AFRICA

433

# INDEX: 2. SOUTH AMERICA

## V

## W

## Y

## Z

# INDEX: 3. SOUTHERN ASIA

## A

# INDEX: 4. SOUTH EAST ASIA

456

# Y

yellowheart  412